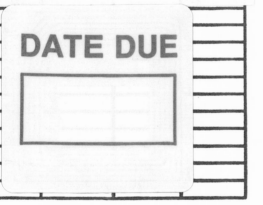

*ENERGETICS OF
PHYSICAL ENVIRONMENT*

ENERGETICS OF PHYSICAL ENVIRONMENT

Energetic approaches to physical geography

Edited by
K. J. GREGORY
Department of Geography, University of Southampton

JOHN WILEY & SONS
Chichester · New York · Brisbane · Toronto · Singapore

Library of Congress Cataloging-in-Publication Data:

Energetics of physical environment.
    Based on lectures presented at the meeting of
Section E of the British Association in Bristol in
Sept. 1986.
    Includes index.
    1. Physical geography — Congresses. 2. Energy budget
(Geophysics) — Congresses. 3. Erosion — Congresses.
4. Bioenergetics — Congresses. I. Gregory, K. J.
(Kenneth John) II. British Association. Section E.
GB3.E54 1987   910'.02   86-26801
ISBN 0 471 91358 8

British Library Cataloguing in Publication Data:

Energetics of physical environment: energetic
    approaches to physical geography.
    1. Physical geography
    I. Gregory, K. J.
    910'.02   GB54.5
ISBN 0 471 91358 8

Printed and bound in Great Britain

List of Contributors

B. W. Atkinson	Professor of Geography, Department of Geography and Earth Science, Queen Mary College, University of London, Mile End Road, London E1 4NS
C. Jane Brandt	Research student, Department of Geography, University of Bristol, Bristol BS8 1SS
K. J. Gregory	Professor of Geography, Department of Geography, University of Southampton, Southampton SO9 5NH
Sheila M. Ross	Lecturer in Geography, Department of Geography, University of Bristol, Bristol BS8 1SS
I. G. Simmons	Professor of Geography, Department of Geography, Science Laboratories, South Road, University of Durham, Durham DH13 5LE
J. B. Thornes	Professor of Geography, Department of Geography, The University of Bristol, Bristol BS8 1SS
D. E. Walling	Professor of Physical Geography, Department of Geography, University of Exeter, Exeter, Devon EX4 4RJ

Contents

K. J. Gregory and J. B. Thornes

Preface

Books on physical geography have been getting longer. During the 1980s there has been a resurgence of attempts to produce physical geography books which cover the entire discipline. Therefore it is perhaps inevitable that physical geography texts have been increasing in size.

This book endeavours to be short and the reason for the book originated from an idea for a day of lectures at the meeting of Section E of the British Association in Bristol in September 1986. The Presidential Address on that day was succeeded by a number of papers following the energetics theme and it seemed timely to collect these papers together and publish them as a single volume. Although that is the reason for this book there is a broader justification. As physical geography texts have become larger they have attempted to cover the whole of the discipline, despite the fact that courses, particularly in Britain, often adopt a particular theme. Students often find that textbooks provide background information, but only a part of their content may be relevant to their courses in physical geography. Therefore, this book endeavours to introduce physical geography from a contemporary viewpoint using a theme which it is suggested may be increasingly pertinent in the next decade. This theme of the power of nature, and of energetics of the physical environment, is one which is in harmony with other branches of science. It is hoped that the introduction given in the subsequent chapters will allow readers to follow the literature in the branches of physical geography, to extend the applications to physical geography as a whole, and to consider other branches of science. Although not intended as a text that would be appropriate to cover an entire physical geography course, it is hoped that this volume could be used as background reading for higher education courses and could be the basis for excursions into the literature.

It is a great pleasure to thank the authors for kindly agreeing to contribute chapters and to producing them promptly; to thank Mr Chris Hill who has done a tremendous amount of support work in relation to development of the theme and preparation of this volume; and not least to Mrs June Gandhi who has coped with the typing of my first chapter and with the preparations for the volume.

KEN J. GREGORY

Energetics of Physical Environment
Edited by K. J. Gregory
© 1987 John Wiley & Sons Ltd

1

The Power of Nature— Energetics in Physical Geography

K. J. GREGORY

Department of Geography, University of Southampton

A power is passing from the earth
To breathless Nature's dark abyss
But when the great and good depart
What is it more than this—

William Wordsworth
Lines on the expected dissolution of Mr Fox

1.1 INTRODUCTION

In these words, written in 1806 at Grasmere, William Wordsworth (1770–1850) acknowledged the power of an individual—Mr Fox. According to Peter Medawar (1986) Sir Francis Bacon (1561–1626) had been associated some two centuries earlier with the notion that the purpose of science is to secure power over nature so that 'human knowledge and power meet in one'. Bacon believed that the purpose of science was to make the world a better place to live in, and according to J. M. Roberts (1980) Bacon advocated the study of nature based upon observation and induction in such a way that it was directed towards harnessing nature for human purposes. Power has thus been associated with individuals and with the effects that human activity can have upon nature. These two ideas have been in existence for more than three centuries, but there is, of course, also a power of nature itself with which this book is primarily concerned.

Physical geography is a comparatively young science and many of its origins appeared during the last century. Although Davies (1968) provided a history of British geomorphology from 1578 to 1878, the first volume of the history of the study of landforms (Chorley, Dunn, and Beckinsale, 1964) began with the contributions of the Greeks but concentrated mainly on contributions made during the 18th and 19th centuries prior to the major contributions of W. M. Davis. More recently, K. J. Tinkler (1985) in his history of geomorphology considered the period before 1800 but emphasized the 18th and 19th centuries prior to indicating selected strands of developments that occurred during the 20th century. Although these three works are concerned primarily with geomorphology, it is true of all branches of physical geography that the antecedents for contemporary approaches emerged largely during the last 100 years. The first half of the 20th century was dominated by an emphasis in physical geography placed upon the evolution of landscapes, and much of this emphasis developed from the work of one man—William Morris Davis (1850–1934)—and from his writings which extended to more than 500 papers and books.

In a review of the nature of physical geography, it was suggested (Gregory, 1985a) that the legacy of such evolutionary approaches still endures in physical geography although it was complemented in the mid 20th century by the impact of the quantitative revolution. Subsequently it has been possible to identify six major themes evident in the endeavours of physical geographers (Gregory, 1985a). The *first* of these themes can be thought of as chronological in that it is concerned with the evolution of the earth's environment and although in some ways it developed from the Davisian approach it was greatly enhanced by the advent of new techniques for chronometric dating. A *second* theme developed as a result of the concentration of research on analysis and understanding of the processes involved in nature.

A *third* theme that became more prominent in the late 1960s and afterwards emphasized the effects that human activity has upon physical environment and upon environmental processes. The need existed for a major approach to physical geography and this proved to be possible, in a *fourth* theme with the advent of the systems approach that was advocated for physical geography as a whole in 1971 by Chorley and Kennedy. In the 1970s there was debate about the extent to which a systems approach could be completely satisfactory as the major underlying paradigm for physical geography, but by the 1980s a number of introductory physical geography textbooks were accepting the systems approach as a fundamental vehicle for an approach to the physical geography of nature (see, for example, Dury, 1981; White, Mottershead, and Harrison, 1984).

In addition to these four themes in physical geography it is possible to discern two others which have appeared more recently. A *fifth* theme may be detected because of renewed interest in the interpretation of temporal change of the physical environment. This renewed interest was catalysed by the inception of new conceptual approaches and it has been made possible because an enhanced understanding of environmental processes can be employed to illuminate the interpretation of environmental change over time. A *sixth* theme has been the greater development of applied physical geography. Although it has always been hoped that physical geography should be useful, the applicable results of physical geography research have tended to emerge incidentally and sometimes accidentally. Most recently, however, physical geographers have acknowledged the part that they can play in providing information as an input to environmental management; the contribution of the geomorphic engineer in geomorphology has been shown to be one that complements contributions made by other disciplines such as civil engineering (Coates, 1976); and it is increasingly recognized that physical geography research must encompass questions of environmental design (Gregory, 1985b).

Physical geography may now have reached the point at which there are indications of advantages of a more unified approach. It is perfectly possible to have six prevailing approaches but it is necessary to concentrate upon the ways in which links can be provided between them. This would not necessarily require the development of a single paradigm but rather would need to concentrate upon links which could be useful both internally in physical geography research and teaching and also externally in the relations within other disciplines. Indeed in such external relations it is paradoxical that the subject of nature and the earth's environment has not captured popular attention in the way that archaeology, geology and history have done. Perhaps the development of a more integrated physical geography could be the catalyst for an understanding of the environment of the earth's surface that would capture more popular interest.

It is suggested that power, that is the rate of doing work, could provide a basic integrating theme. Such a theme should be capable of providing a unified approach to nature or to physical environment and it should also be capable of extending the existing six approaches referred to above. Thus in the study of processes, which have tended to be conducted rather separately in the branches of physical geography concerned with the atmosphere, hydrosphere, geosphere, pedosphere, and biosphere, it is desirable to adopt an approach which is consistent for each of the spheres and also is capable of extending investigations of process beyond a mere budget or energy balance. To focus upon power in investigations of process may therefore deflect the tendency towards a more realist approach (Chorley, 1978) which has the disadvantage that it leads towards the physical sciences and takes physical geography away from the mesoscale focus of attention that it should properly adopt. In the adoption of the systems approach in physical geography the major emphasis has been placed upon entitation and quantitation (Huggett, 1980) and perhaps insufficient emphasis has been devoted to system dynamics. Therefore the major achievement of the systems approach to date has been to describe the structure of systems in nature rather than focusing sufficiently upon the energy flows which link the structural components and which therefore depend upon the power distribution in particular systems. The possible integrating role of the power theme can also apply to the other themes recently evident in physical geography. Thus studies concerned with change over time basically depend upon the ways in which power expenditure in nature varies over time and upon the way in which such changes can be dated, modelled and explained. Many changes in power distribution and expenditure can be ascribed to the effect of human activity and in the field of applied physical geography it is often the case that the optimum way of managing the earth's environment is to minimize the impact of disruption and therefore to minimize power expenditure.

In addition to extending, and to some extent unifying, the recent approaches to physical geography, it is possible that a focus upon power could relate to a number of outstanding questions which confront physical geography. It is necessary to consider whether power and energetics as reviewed below could provide a single underlying paradigm. In human geography Johnston (1979) concluded that the current approach was founded upon multiple paradigms. However, for an approach to the physical environment the power of nature can provide a fundamental basis. It has also been questioned how far physical geography can maintain its mesoscale focus and how necessary it is for the subdivisions of physical geography to adopt a realist approach. A further problem that has been acknowledged is to question whether physical geography is internally unbalanced because of the dominance of geomorphology and disintegrating so that Brown (1975) was concerned that physical geography was in the process of being rediscovered under the guise of environmental science.

Are we therefore approaching the time at which it is appropriate to contemplate a more integrated physical geography and can such an integration be achieved based upon power and energetics? This book is devoted to an exploration of the theme of the power of nature and to consideration of this theme as a primary focus for physical geography. This is approached in the remainder of this chapter by tracing the recent development of the concept in physical geography (1.2); by comparing the situation with other disciplines (1.3); by proposing an outline for an energetic physical geography (1.4); and by referring to an example of energetics as illustrated by the fluvial system (1.5).

1.2   ENERGY IN PHYSICAL GEOGRAPHY

Although considerable scope remains for further development of a unified approach to power, energy and therefore to energetics of the whole physical environment, a number of important contributions have already been made. Figure 1.1 endeavours to summarize contributions in books and research articles by a range of physical geographers. For any specific contribution to be included the approach had to refer explicitly to the need to concentrate upon energy, power or rate of doing work in the physical environment.

The trend line (Figure 1.1) was constructed simply to reflect a cumulative total of contributions and it can be seen that since 1953 the number has increased quite significantly, and particularly in the 1970s and 1980s the number of contributions using an energetic approach to physical geography has been considerable.

It is possible to distinguish those contributions that are general and apply to either the whole of physical geography or to large parts of the field, from those contributions that have been made specifically in the context of the branches of physical geography, namely meteorology and climatology, geomorphology, pedology, biogeography, and more recently in relation to hydrology. As increasing interest in energetic physical geography has arisen over the last three or four decades, there has been a significant influence by developments in other disciplines, and in one of several important papers in *Geography* in 1965 Stoddart (1965) notes the way in which the energy system can link geography to the mainstream of scientific thought (see Section 1.3). However, in Figure 1.1 the emphasis is placed upon contributions written by physical geographers although the definition of physical geography becomes less clear in North America and particularly in the United States.

When looking at the way in which energetic physical geography has developed, it is in the field of meteorology and climatology that the approach was first clearly evident. In his book *The Restless Atmosphere*, F. K. Hare (1953) clearly suggested that a geographical approach to the atmosphere should be founded upon the energy balance. The importance of this approach in climatology has been continued in a number of subsequent studies, for example by Budyko (1956) and in subsequent papers by F. K. Hare. In 1965 Hare emphasized energy exchanges within the atmosphere where energy signifies the capacity to do work, and in 1966 Hare proposed that the outstanding change in the study of climatology since 1945 had been the shift away from the measurement of parameters such as temperature and relative humidity towards the measurement of fluxes. This change meant that the concern with the movement and transformation of energy in the atmospheric boundary layer also extended to embrace the plant cover and the soil so that progress was made towards an understanding of the mechanism of energy and moisture exchange. An important subsequent feature of research in climatology was the emphasis upon establishing energy budgets for specific areas. In a review of energy budget climatology (Hare, 1973), it was suggested that the contributions that have been made are analogous to the energy-centred methods that have been adopted by ecologists. More recently a view of climatology for geographers proposed by Terjung (1976) advocated the study of flows of energy, mass, momentum and information through the various environments of the planet earth. Terjung concluded that geographers interested in the physical environment of man need to be trained rather differently in the future than they were in the past. He argued that prospective physical geographers will need to take basic courses in calculus, physics, chemistry, engineering, modern biology and computer programming. Other climatologists have also recognized the value of an energetic approach, and Mather *et al.* (1980) argued that climatology must systematically investigate the exchanges of heat, water and momentum that occur at or near the earth surface. In dynamical meteorology Atkinson (1981) also recognized the importance of an energetic approach to the atmosphere.

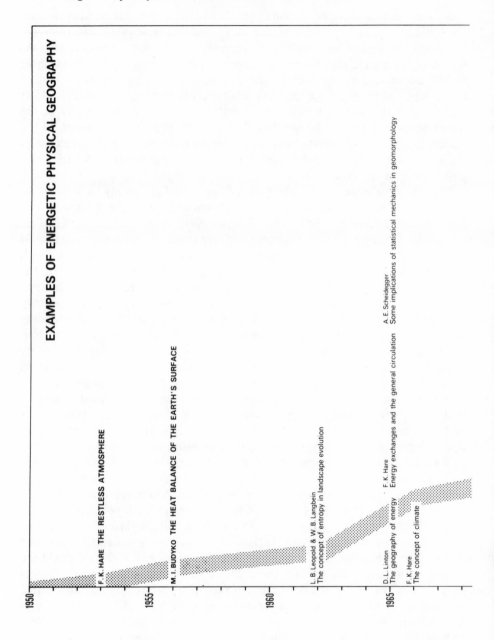

EXAMPLES OF ENERGETIC PHYSICAL GEOGRAPHY

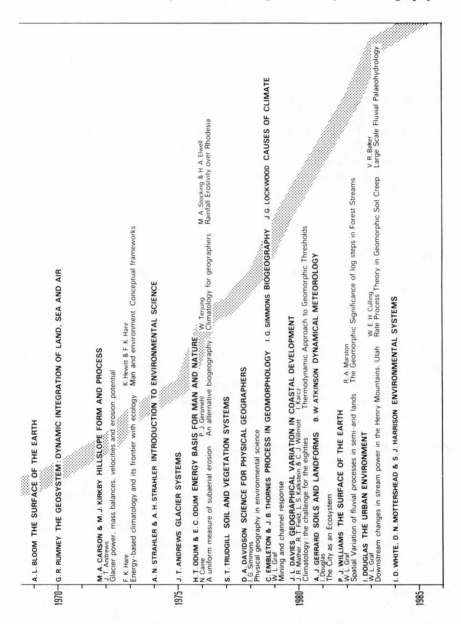

Figure 1.1 The use of energetic approaches in physical geography publications. Books are shown in upper case, journal articles in lower case, and the trend line shows the cumulative total of publications cited

In biogeography the importance of an energetic approach has been acknowledged more recently than in climatology although ecologists have recognized the significance of such an approach for some time. The study of energy in contemporary society was advocated by Simmons (1978) where he adopted the taxonomy of energy systems employed by ecologists and proposed that geographers could use a set of ecosystem types which broadly provide a set of spatial regions which conform to patterns identified from analysis of satellite data. Subsequently in his text on biogeography Simmons (1979) used energy as a key for the understanding of natural biogeography because food chains, productivity, nutrient cycles, and population dynamics could provide the basis for his subsequent treatment of cultural biogeography. In the study of soils, although an energy-based approach has been developed by pedologists recently, soil geographers have also recognized the value of an energetic approach, and Gerrard (1981) reviewed the energy status of soil systems in relation to the decay component, the cyclic component and the random component.

In geomorphology some of the most intriguing contributions have been made in the context of specific problems. In 1962 the relevance of entropy in landscape evolution was explored by Leopold and Langbein. Subsequently in 1972 a very imaginative approach was devised by Andrews when he provided an analysis of total glacial power (WT) as the product of basal shear stress and the average velocity. Effective power (WE) was determined by the proportion of the total average velocity which resulted from basal sliding so that the ratio between total glacial power and effective power (WT/WE) would vary according to the proportion of basal slip to internal ice deformation. This approach to glacier power enabled Andrews (1972) to draw distinctions between the glacial erosional forms produced by arctic and temperate glaciers. A further very important application in geomorphology was made by Caine (1976) when he was able to estimate the physical work in joules represented by different types of sediment movement for areas in Colorado and Sweden. In an important paper in 1979 W. L. Graf proposed an approach for the investigation of mining and channel response. He concluded that the tradeoff between force and resistance lies at the heart of explanation in geomorphology. He developed methods to deduce tractive force and to express resistance by estimates of biomass and showed how the relation between the two changed during the 19th century and had an effect upon the fluvial system. In subsequent research Graf (1982, 1983) has explored spatial variations of fluvial processes and has shown how spatial patterns of stream power change from one time period to another (see p. 24). Such papers provide recent examples of very stimulating geomorphological contributions using energetics and power and they offer an approach with considerable future potential. Further examples are referred to in Section 1.5 below.

In the branches of physical geography there are clear indications of approaches based upon energy, power and energetics. In addition, however, there are suggestions that such an approach can be adopted more widely in physical geography and this has been evident in a number of recent textbooks. However, one of the earliest approaches of this kind was in a paper by Linton (1965) on the geography of energy. Taking geography to be the description of the changes that take place or have taken place in or at the surface of the earth, Linton (1965) proposed that any changes which occur in the real world imply that work has been done and energy has been expended. He identified four sources of energy: radiant energy from the sun; internal energy from the earth's interior; rotational energy of the whole and parts of the solar system; and vital energy which is energy in the service of man. Linton concluded with the hope that his method of expressing salient parameters in terms of a common set of units, the watt and the calorie, has value for the future of geography. Although Chorley (1973, p. 157) argued that the geography of energy could not easily apply to human activity and particularly to the mechanisms of group decision-making; nevertheless the approach suggested by Linton is one that is attractive and commendable for physical geography. Subsequently, the energy approach has appeared to varying degrees in other textbooks including those by Bloom (1969), Rumney (1970), and Strahler and Strahler (1984).

Other general approaches have appeared in relation to process in geomorphology, and Embleton and Thornes (1979) visualized energy as attributed to solar radiation, atomic energy, chemical energy, gravity and energy of earth's rotation sources and introduced these energy types as fundamental to the understanding of geomorphological processes. Although concerned with only a part of geomorphology, the book by P. J. Williams (1982) entitled *The Surface of the Earth; An Introduction to Geotechnical Science*, is one which adopts an approach clearly based upon energy. Williams argues (1982, p. 20) that

> the materials of the earth's surface behave in accordance with fundamental principles relating to energy and matter that have been thoroughly established by the work of physicists, chemists, and others over several centuries. Interpretation of geomorphic processes requires that we apply such principles with precision and a proper degree of understanding of their meaning. Failure to do so results in theories or dogma which are at best meaningless and at worst a source of danger to man's well being.

In his text Williams (1982) then proceeds to suggest how equilibrium thermodynamics, chemical thermodynamics and energy balance can be the basis for an approach to the geomorphological study of the surface of the earth. A number of other textbooks have adopted an energy basis (see, for example, White, Mottershead, and Harrison, 1984) although sometimes the approach advocated in the introduction is not followed throughout the book as completely as it is in the text by Williams (1982).

There are two other general ways in which an energetic approach has appeared in the writings of physical geographers. In the case of a specific type of area, Douglas (1983) has developed an ecosystem view of the city. He considers the energy balance, the water balance, mass balance geomorphology, biogeography and waste disposal of the city as a basis for looking at geographical aspects of urban health and disease and then at management and planning design to reduce environmental hazards. In his novel approach to urban physical geography Douglas (1983) is emphasizing the flows of energy through all the aspects of the urban ecosystem. A further approach is adopted by D. A. Davidson (1978) in his book *Science for Physical Geographers*. His basic premise is that there is a body of scientific information which is a prerequisite for the physical geographer and a significant amount of this basic information depends upon energy and energy transfers.

There is therefore ample evidence that during the last 30 years physical geography both in general and in its component branches has been moving towards a more energy-based approach. Sufficient examples are collected together in Figure 1.1 to indicate that a focus upon energy and energy transfer may therefore provide an important fundamental theme for physical geography as a whole and for the branches of the subject. To develop such a fundamental basic approach is not necessarily easy of course, and when Trudgill (1977) wrote on soil and vegetation systems he stated that his initial intention was to deal in detail with the three main systems of flow and cycling nutrients, energy and water. Although he did not follow this initial intention the energy flows are prominent in his book. Perhaps the most generally applicable energy-based approach was advocated by Hewitt and Hare (1973). They proceeded from a review of exchange of energy and mass in the biosphere and between the atmosphere and the landsurface, and they showed how the functions of an ecosystem require an unending series of exchanges of energy, water, atmospheric gases and mineral nutrients between the organic and inorganic parts of the system. Although models of paths and storage reservoirs have been developed by ecologists, geochemists and climatologists they proposed that there was a need to progress towards a multidimensional model of the entire system. In reviewing conceptual approaches to man and environment, Hewitt and Hare (1973) acknowledged not only the fundamental importance of an energy-based approach but also the need to appreciate developments in related disciplines. It is to such developments in other disciplines that attention must be turned prior to suggesting the architecture for an energetic physical geography.

1.3   ENERGY AND ENERGETICS IN SCIENCE

Until the last decade it is curious that physical geography appeared to be unaware of conceptual developments in science. Since 1969 when David Harvey published *Explanation in Geography*, this deficiency has begun to be rectified. Particularly since 1975 there has been acknowledgement of the need to be aware of important conceptual progress in science and this is noted by Gregory (1985a) and developed by Haines-Young and Petch (1986). Developments in science which provide a context for an approach in physical geography can be seen in four ways. First the origins of energy and energy related concepts in science (1.3.1); secondly, developments in fields related to physical geography especially ecology (1.3.2); thirdly the development of new disciplines (1.3.3); and fourthly an appraisal of recent challenging developments in science as a whole (1.3.4).

1.3.1 Related concepts in science

A review of the historical development of the concept of energy has been provided by Lindsay (1976). He noted that Leibniz in 1695 used terms relating to dead and live forces and these terms were subsequently to lead to the formalization of potential and kinetic energy. Not until 1830, however, did Hamilton formalize the definition of kinetic and potential energy and then $PE + KE$ was recognized as the Hamiltonian, mechanical energy. Sadi Carnot in the early 19th century implicitly recognized work, as defined by force times distance, and appreciated that heat had motive power in his development of heat engines. Scientists such as Joule in 1847 and Mayer showed that heat should be treated as a form of energy into which mechanical energy could be transformed and vice versa. Carnot's work laid the basis for thermodynamics and particularly for the second law. Pressure, volume, chemical composition and temperature are the classical physicochemical parameters in terms of which the properties of macroscopic systems are defined (Prigogene and Stengers, 1984), and thermodynamics is the science of the correlation among the variations in these properties. In 1847 Von Helmholtz introduced the law of conservation of energy which came to be known as the first law of thermodynamics. In 1852 William Thompson formulated the second law of thermodynamics which related to the irreversibility of energy dissipation. This introduced the notion of time, and in 1865 Clausius introduced the concept of entropy. Entropy, concerned with the distribution of energy and the probability of its distribution at a particular time, naturally furthered the notion of time in physical science so that Eddington (1958) regarded entropy as the 'arrow of time' or an 'indicator of evolution'. Because thermodynamics was concerned with the qualitative diversity of energy and its tendency towards dissipation, it was necessary to consider how entropy could be equated with increasing molecular disorder, and Boltzmann introduced his order principle which considered the probability of entropy of a particular system.

These developments in the 19th century were acknowledged in the study of environment in the mid 20th century. In a book entitled *Energy Basis for Man and Nature* H. T. and E. C. Odum (1976) began with the precept that everything is based upon energy and that energy flows in ecosystems embrace both cultural and natural components. In their book, in addition to clear definitions of energy, power as the rate at which energy flows, and efficiency, which is any ratio of energy flows, they propose three principles of energy flows. These were: the law of conservation of energy; the law of degradation of energy, which involves entropy as a measure of technical disorder to signify the extent to which energy is unable to do work; and the principle that systems which use energy best are the ones which survive and this is the maximum power principle or the minimum energy expenditure principle. Although applied to man and nature in general, their book owes much to developments in ecology which provide a second strand for consideration of developments in science.

1.3.2   Developments in ecology

In ecology the recognition of trophic levels by Lindeman (1942) established that living organisms can be separated into a series of more or less discrete levels with each trophic level depending on the preceding one for its energy supply. Therefore, this approach to ecology and the ecosystem involved documentation of energy flows within the ecosystem and required elucidation of trophic structure and of energy flows within different types of ecosystems. A historical review of energy flows in ecosystems was provided by Odum (1968) who suggested a general energy flow diagram. In some ways analogous to the flow of energy in ecosystems is the movement of nutrients in ecosystems and in specific examples it has been possible to suggest the main pathways of nutrient cycling which occur. Subsequently an important development has been the focus upon productivity because the ecological productivity is determined by the amount of solar radiation available to primary producers at trophic level 1 for photosynthesis and also the efficiency with which autotrophs convert solar energy into usable forms by photosynthesis. A general review of energy in biogeographical systems and an outline of productivity is provided by Park (1980), and the progress is more completely explained in Chapter 6 (p. 146).

1.3.3   New disciplines

In science in general a concern with energy has been reflected in the way in which new disciplines have been advocated (Table 1.1). Following the realization of the implications of thermodynamics for the development of systems approaches, the field of cybernetics was suggested in 1948 by N. Wiener to refer to control and communication in the animal and the machine. This field of study was primarily concerned with control mechanisms in systems and with communication processes which determine the successful operation of those systems and part of its mathematical basis is found in information theory. Although *cybernetics* was initially conceived as a study of regulating and self-regulating mechanisms in nature and in technology, it subsequently became

Table 1.1   Some new disciplines related to energetics

	Trophic dynamics
1942 R. L. Lindeman	. . . the transfer of energy from one part of the system to another
	Cybernetics
1948 N. Wiener	. . . control and communication in the animal and the machine
	Geocybernetics
1963 M. L. Polonskiy	(cited by D. R. Stoddart, 1967)
	Energetics
1966 J. Phillipson	. . . concerned with the energy transformations which occur within ecosystems
	Geosystem
1970 G. R. Rumney	Dynamic integration of land, sea and air
	Energy in thermodynamics
1971 E. P. Odum	The one-way flow of energy, as a universal phenomenon in nature, is the result of operation of the laws of thermodynamics, which are the fundamental concepts of physics
	Bioenergetics
1975 E. Broda	. . . for an understanding of more and more complicated systems, thermodynamics and kinetics must increasingly be supplemented by cybernetics, by applied systems analysis.
	Synergetics
1985 H. Haken	. . . the cooperation of individual parts of a system that produces macroscopic spatial, temporal and functional structures.

associated particularly with negative feedback systems. Thus Coffey (1981) referred to a field of 'second cybernetics' being introduced where the concern with feedback is emphasized. Physical geography has acknowledged the existence of the field of cybernetics but has made little conscious attempt to adapt the field, although Stoddart (1965) referred to a little-known paper where Polonskiy in 1963 suggested a science of geocybernetics presumably concerned with control and communication in the geosystem.

Such developments therefore embrace the field of *energetics* which was defined in 1966 by J. Phillipson as a field of investigation 'concerned with the energy transformations which occur within ecosystems'. Until the implications of the first and second laws of thermodynamics had been fully assimilated, progress in the analysis of bioenergetical processes was uneven and patchy (Broda, 1975). Recognition of a field of *bioenergetics* arises because understanding of more and more complicated systems requires that thermodynamics and kinetics must increasingly be supplemented by cybernetics, and Broda (1975) regarded organisms as chemodynamical machines and identified three classes of bioenergetic processes: fermentation, photosynthesis and respiration. The field of biognergetics has been advocated by Ji (1985) who recommended the term to connote the study of living processes from the point of view of energy and information.

Although emphasizing that we should never forget the limitations of universal approaches, Haken (1979) contemplated whether there are unifying principles in science. He recognized (Haken, 1979) *synergetics* as an interdisciplinary field of research that is concerned with the cooperation of individual parts of a system that produces macroscopic, spatial, temporal and functional structures. This field of synergetics deals with deterministic as well as stochastic processes, and although the field originated in physics links exist with other branches of investigation including bifurcation theory in mathematics, the study of lasers, of fluids, studies in chemistry, biology and even in sociology. This field of synergetics, because it embraces concepts of order parameters, instability and slaving is suggested to hold further promise for application in fields of geography (Haken, 1985).

Such new fields of cybernetics, bioenergetics and synergetics, show that in recent years new conceptual approaches have been evident in several fields of science and such approaches could have significant implications for physical geography. It has been suggested that whereas general systems theory, thermodynamics and statistical mechanics held promise during the past decade, new concepts hold promise for geomorphology and other branches of physical geography at the present time and perhaps into the next decade. Thus Karcz (1980) has indicated that we are no longer concerned simply with whether a system tends to a steady state, but also whether the steady state is stable and persists, or whether it is unstable and evolves to a new temporary steady state possibly with a greater degree of coherence. Whereas the 'thermodynamic branch' deals with stable steady states and self-stabilizing feedback regimes, it is possible that beyond that domain orderly configurations arise and systems may evolve through series of successive, instability-triggered transitions leading to a more complex organized state. Such transitions are associated with increased entropy production and are characteristic of non-linear regimes far removed from equilibrium (Karcz, 1980).

1.3.4 Contemporary science

It is difficult in a few sentences to even introduce the new developments in science which are stimulating approaches to physical environment. It has been argued that there is a need for unification as a reversal of the trend towards fragmentation of human knowledge, and to this end Bohm (1980) has distinguished explicate order, as the order we commonly encounter and experience, from the implicate order which is the enfolded order that is completely embraced within nature. Bohm has further suggested that physical laws have so far referred to explicate order and need to be transferred to dependence upon implicate order. He speculates further (Bohm, 1980) that we may need to combine non-equilibrium thermodynamics with catastrophe theory, so that this could provide real understanding of morphogenesis in physics and in nature. Prigogine, who won the Nobel Prize in 1977 for his investigations of the thermodynamics of non-equilibrium systems, has argued that we need to concentrate upon what he describes as dissipative structures which arise out of non-linear processes in non-equilibrium systems. Implicit in this development is the relationship between function, structure, and fluctuation which Prigogine expressed in the form

$$\text{Function} \rightleftarrows \text{Structure}$$
$$\searrow \quad \swarrow \quad$$
$$\text{Fluctuation}$$

Other significant developments have emerged from particular areas of science. In the context of better chemical process design, Linnhoff (1983) defined energy as the ability to deliver work. To organize the bewildering collection of existing experimental data in cell biology, Ji (1985) has envisaged similarities between the algebra of machines, information theory and molecular theoretic approaches which may eventually allow existing experimental data to be organized into a coherent body of knowledge similar to the periodic table. Such developments take consideration of energy beyond a simple energy budget approach, and Haigh (1985) has explained organicism as the investigation of the complexity of reality which can contain components that have the ability to become organized into entities with new properties. He suggests that general systems theory represents an exciting and challenging routeway not just to the reunification of geography but to a new unified understanding of the unity of nature, science and society.

The existence of new exciting approaches in different branches of science and the definition of new fields of enquiry both indicate that to focus upon energetics of the physical environment and therefore upon the power of nature is timely and viable. In at least one recent text the energetics approach has been employed as a way of regarding the universe, exemplifying the range of spatial and temporal scales that its investigation demands. Thus Taube (1985) reviewed energy and nature, including natural sources, the energy needs of man in a developed society, and the technological use of energy. His book involved expressing a wide variety of energy sources and energy transfers in consistent quantitative terms, and as such makes considerable progress in this field.

1.4 ENERGETIC PHYSICAL GEOGRAPHY

Sufficient examples have been given above to show that there is now ample opportunity for a more unified approach to energy and energy transfers in the physical environment. Such an approach can be stimulating as in the case of the study by R. J. Blong (1982) of *The Time of Darkness* in which he considered the thermal energy production of the Tibito Tephra eruption that occurred in Papua New Guinea, and which qualifies as one of the greatest eruptions of the last 1000 years. In branches of physical geography such as those including the study of volcanoes, it is an obvious approach to compare the energetics of specific eruptions (see, for example, Blong, 1984). However, is it possible to adopt a similar approach throughout all branches of physical geography? Many studies previously undertaken (Figure 1.1) have involved reference to particular aspects of energy and power in the physical environment. In Figure 1.2 there is an indication of the major themes which were identified as integral components of an energy-based approach in a large number of the books and papers which go to make up Figure 1.1.

It is evident that although the major themes identified have been concerned with sources, with systems and with transfers of energy and also with entropy, nevertheless other aspects have featured in addition. Such energetic foundations for physical geography have not been confined simply to the traditional divisions of physical geography, namely those concerned with the atmosphere, the geosphere, the pedosphere, the biosphere and more recently the hydrosphere. In his analysis of the urban environment, Douglas (1983) laid considerable emphasis upon the energy balance of the city, upon the energy balance as modified by human activity, and also extended this to a comparison of natural with people-made energy flows. The way in which the energy balance in the context of urban areas was then employed (Douglas, 1983) to illuminate the geomorphology and biogeography of the city and aspects of city management, demonstrate how an energetic approach can usefully provide a very significant foundation for physical geography.

Energy can be envisaged as the capacity for doing work and as a basis for his introduction to the climatic system; for example Lockwood (1979) refers to the variety of forms of energy including, heat, radiation, potential energy, kinetic energy, chemical energy and electric and magnetic energies. It is also necessary to consider work and power defined as follows:

Work   Force × distance   joule (J) — work done when a force of 1 newton (N) acts over a distance of 1 m

Power   Work done per   watt — power expenditure of 1 joule (J) per second
unit time

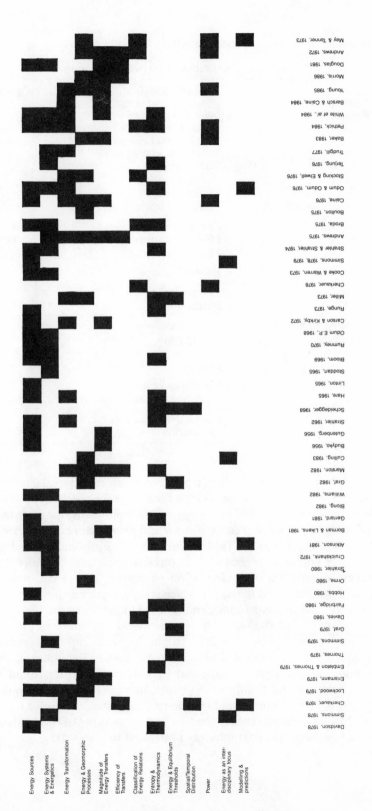

Figure 1.2 Major themes evident in published examples of energetic physical geography. The context of 57 publications, many of which are listed in Figure 1.1, is analysed according to 13 topics listed on the left-hand side

Table 1.2   Investigations in physical geography

Energy phase	Type of identification in physical geography
1 Energy sources	Solar energy
	Geothermal energy
	Gravitational energy
	Rotational energy
	Vital energy
	Potential energy
	Kinetic energy
	Enthalpy (free energy)
2 Energy circulation transformation transfers	Conservation of energy
	First law of thermodynamics
	Fluxes, energetics
3 Energy budget	Energy balance
Availability	Thresholds
	Entropy
4 Energy related to morphology (rate of doing work)	Equilibrium
	Steady state
	Power
	Efficiency
5 Changes of energy distribution	Maximum power
	Minimum variance
	Dissipative structures
	Fluctuation

The amount of energy available in a particular part of physical environment will influence the work that can be done and therefore the potential power that is available. It is therefore possible to suggest, very simplistically, the ways in which aspects of energy, and subsequently of work and power, can be identified in the fields of physical geography. In Table 1.2 are listed five aspects of investigations which have been applied to several fields of physical geography. The aspects range from energy sources, through circulation of energy, to the construction of budgets and subsequently to the way in which energy, work and power are related to morphology, and finally to consideration of changes over time. This involves not only explanation of the short-term adjustments, but involves longer-term considerations as well and may lead to the inclusion of consideration of dissipative structures as indicated in Section 1.3 above. Examples of the five major groups of investigation indicated in Table 1.3 will be found in the subsequent chapters and finally in the conclusion (p. 161) to this book there is a further table which illustrates how the five suggested categories have been employed. In addition, reference to the first and second laws of thermodynamics has featured in physical geography, as illustrated in Table 1.3.

Table 1.3 Examples of definitions of energy dynamics in relation to physical geography

---

First law of thermodynamics

Thermodynamics means the study of heat as it does work, but this narrow definition is rather misleading, perhaps thermodynamics could more aptly be termed energetics as it is now concerned not only with heat, but also with all other forms of energy.

White, Mottershead and Harrison (1984, p. 9)

The first law of thermodynamics expresses the conservation of energy whilst the second law is concerned with transfer of energy from one state to another.

Davidson (1978)

Energy may be transformed from one form into another but is neither created nor destroyed.

Phillipson (1966, p. 3)

Second law of thermodynamics

The second law of thermodynamics is concerned with the direction of naturally occurring or real processes.

White *et al.* (1984, p. 32)

The direction of change in a system is dictated by the tendency of the system to achieve a lower energy state and also to reach a more uniform configuration.

Davidson (1978, p. 58)

The second law specifies that any system without inner constraint tends over time toward an irrevocable state of equilibrium or perfect homogeneity (disorder) in which it is unable to produce work.
Coffey (1981, p. 203)

Entropy

The level of entropy—i.e. of homogeneity, randomness or lack of free energy—thus uniquely describes the state of a bounded system.

Bennett and Chorley (1978, p. 5)

Used very broadly, it is a measure of the randomness of a spatial organisation—the probability of encountering given states, events or energy levels throughout the system.

Chorley and Kennedy (1971, p. 348)

Entropy is a measure of the disorder of the system, but it can never be absolutely quantified. However, it always increases with the operation of natural processes.

White *et al.* (1984, p. 33)

Another way to describe the tendency for potential energy to be used up and degraded is to say that entropy always increases in real processes.

Odum and Odum (1976, p. 38)

We should therefore consider whether it is possible to express work in terms of joules (J) and power in terms of watts (W) for all aspects of environmental systems. Such expression could provide a useful standardized way of envisaging energy in environmental systems as illustrated by the use of glacier power (Andrews, 1972), the use of power expressed in watts for slope and stream

Table 1.4   Work in specific aspects of physical environment

World solar radiation	15.5×10^{21} J day^{-1}	3 263 158 ×
World internal energy lost	2.74×10^{18} J day^{-1}	5768 ×
Volcanic eruptions		
Krakatoa, Indonesia (1883)	1.0×10^{18} J	2105 ×
Mount St Helens, USA (1980)	2×10^{16} J	105 ×
Landslides		
Flims, Grisons, Switzerland (10 000 BC)	3.5×10^{17} J	737 ×
Huascaran, Peru (1970)	2×10^{15} J	4.2 ×
Thermal waste power of world		
1970	4.75×10^{14} J day^{-1}	1
2000	27.48×10^{14} J day^{-1}	5.8 ×

processes as effectively exemplified by Caine (1976), and energy used as the basis for comparison of different volcanic eruptions by Blong (1984). In specific cases the expression of particular aspects of environmental processes or of specific events in terms of energy can provide a useful way of appreciating the dimensions of the power of nature. Thus in the case of the landslide which occurred in 10 000 BC at Flims, Grisons, Switzerland it was suggested (Erismann, 1979) that the energy involved in the landslide would suffice either to cover the total world's energy consumption during 10 hours, or would be sufficient to accelerate the Cheops pyramid into a satellite orbit. The Kofels landslide in the Oetztal, Tyrol of Austria which occurred about 6700 BC was 20 times smaller than the Flims slide, but the energy involved would be sufficient to completely melt a granite pyramid of the same size as Cheops. Such values indicate the utility of comparing different amounts of work done, and in Table 1.4 there is a simple comparison including the Flims landslide and also including other aspects of work done in physical environment all shown as multiples of the thermal waste power of the world based on the figure estimated in 1970. Such comparisons involve a considerable number of assumptions but nevertheless give ideas about the order of magnitude of energy available and work done. In his book considering energy on a universal scale Taube (1985) has endeavoured to standardize energy for the universal scale and has compared energy appropriate to a whole range of different scales from particle to planet.

In physical environment it is not always easy to proceed to obtain estimates for work and power because work requires inclusion of distance and therefore of spatial dimensions, and power involves consideration of time so that the temporal dimension is the basis for calculating work done per unit of time. However, to provide estimates of world energy flux expressed in watts it has been possible to construct Figure 1.3 which endeavours to demonstrate the comparative rates at which energy flux directs physical environment. Considering the logarithmic scale for the depiction of energy flux in Figure 1.3, it is apparent that there are great orders of magnitude difference between the components.

GLOBAL ESTIMATES OF ENERGY FLUX

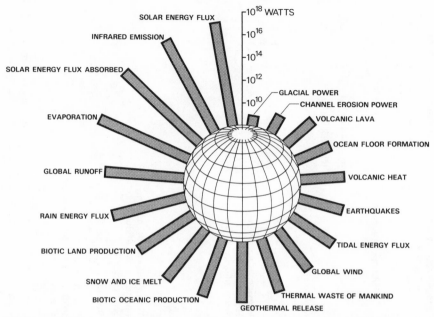

Figure 1.3 World energy flux. A diagrammatic attempt to indicate the relative significance of major fluxes in the earth's physical environment. The scale is logarithmic and the figure summarizes the absolute power of nature over the earth's surface

The importance of solar energy is emphasized, but it is worth recalling that the world use of energy in 1970 (Steinhart and Steinhart, 1974) is 1/3600th of the total solar energy flux and therefore not so far displaced from the magnitude of incoming solar radiation.

Whereas in Figure 1.3 the estimates of world energy flux are given as total power over the globe, it is also necessary for specific processes to think in terms of the power expenditure per unit area, and this is attempted in Table 1.5.

This allows comparison of major river floods with the runoff intensity expressed in stream power, hillslope processes and rainfall power all expressed in watts per unit area. Although this shows (see Table 1.5) how the unit power is greatest in the context of major floods like the Missoula flood arising from the sudden discharge of water from glacial Lake Missoula with maximum water discharges that have been estimated as 38 cubic kilometres of water per hour, nevertheless this is a useful way to compare precipitation power with runoff, and slope process. Surprising contrasts in the power of events may be revealed by comparisons based upon power expenditure per unit area and intensity of action and absolute size need to be considered (Baker, 1983). Although power has been used in relation to atmospheric systems for many years and the

Table 1.5 Examples of power expenditure per unit area in physical environment to contrast with total power estimates as shown in Fig. 1.3

POWER/UNIT AREA Wm⁻²

	Rainfall power		Stream power	Hillslope processes	Rivers in flood
	Intensity	Median diameter	Runoff intensity		discharge (m² s⁻¹)
10^5					
10^4					Missoula Flood 1×10^7
10^3					Elm creek 1×10^3
10^2					
10^1	1 10 cm/h⁻¹ Cloudburst	2.55 mm		Heaving (temperature) Heaving (frost) Water (dissolved)	Amazon, Mississippi East Fork (Wyoming) 3×10^5 3×10^4 23
10^{-1}	10^{-1} 1.5 cm/h⁻¹ Heavy rain	2.05 mm	80 mm/h⁻¹		
10^{-2}	10^{-2} 0.1 cm/h⁻¹ Light rain	1.24 mm			
10^{-3}			0.2 mm/h⁻¹		
10^{-4}				Water (suspended), heaving (moisture) Mudflows Rain impact Gravity	
10^{-5}				Rockfall Surficial wasting	
10^{-6}				Avalanche	
10^{-7}				Soil creep and solifluction Solute transport	
10^{-8}				Gravity	
	Source: R. Lal (1977)		Source: J. B. Thornes (1979)	Source: M. A. Carson and M. J. Kirkby (1972) N. Caine (1976)	Source: V. R. Baker (1983)

production of energy budgets has been an integral part of climatology (Hare, 1973) nevertheless an important use of power has been found in relation to analysis of the fluvial system and this is outlined briefly in the following section.

1.5 POWER IN THE FLUVIAL SYSTEM

Stream power was introduced by Bagnold (1960), and defined as the product of fluid density (p), slope (s) acceleration due to gravity (g) and discharge (Q). As such the definition of power can of course be applied to any fluid, and Bagnold used a similar approach in relation to wind movement over the earth's surface. A stream was conceptualized by Bagnold (1977) as a transporting machine whose efficiency in using available power can be applied to the rate of sediment transport. In a river channel water performs work by overcoming internal friction (viscous shear and turbulence), by overcoming friction at the channel boundary, by eroding the channel boundaries, and in transporting the sediment load (see also pp. 77–79).

Recent research has developed the notion of stream power ω expressed as:

$$\omega = pQgs$$

in several ways (Gregory, 1982), and at least four can be summarized here.

First, and related to consideration of sediment transport, it has been possible to express fluvial processes in terms of stream power. Although research results have often been related to stream discharge, it can be argued that whereas discharge is essentially a hydrological expression stream power is an expression more appropriate for geomorphological analyses. Sediment transport equations have been devised in terms of stream power (Allen, 1977).

Therefore it has been important to see the way in which stream power has been estimated for rivers of different kinds, and maps can be produced to show the spatial variations which exist; Ferguson (1981) provided maps showing stream power values for a number of rivers in Britain (Figure 3.10), and Lewin (1983) showed how stream power values for rivers in central Wales could be related to rates of bank erosion.

Surface runoff can be expressed in terms of kinetic energy for specific rainfall events, and a much higher percentage of rainfall is converted into surface water kinetic energy in an urban than in a rural watershed (Cherkauer, 1978). Conversely baseflow kinetic energies were much higher under rural than under urban conditions. Cherkauer (1978) proceeded to conclude that a very substantial modification of the energy—magnitude frequency relationships had accompanied urbanization, and that future investigations of the impact of urbanization could be conducted through a study of the energy relations and their modification by urban development.

A *second* theme has been to focus upon the power budget and this leads to consideration of power efficiency. Thus Bull (1979) compared the power needed to transport the sediment available at a particular point along the stream channel with the power actually available. If the two values were identical then the situation was described as critical power; but if more power was available than was necessary then erosion could occur, whereas if less power was available than required to transport the sediment supplied then aggradation may occur.

A *third* way in which power has been used in fluvial systems is by establishing relations between power and morphology aspects of the fluvial system. Thus the network of a drainage basin could be visualized as lying on the scale between maximum efficiency at one extreme and minimum power expenditure at the other (Leopold and Langbein, 1962; Chorley and Kennedy, 1971). To devise a way of expressing the volumetric nature of the stream channel system, an index of drainage network power was suggested and discussed (Gregory, 1979; Knighton, 1980). In approaches relating power to morphology it has been usual to use unit stream power as initially defined by Yang (1971). Such unit power is obtained by relation of power available to a unit measure which is usually unit channel width. It has been shown that the way in which unit stream power expenditure varies can be employed to explain characteristics of the long profile, stream channel pattern, the hydraulic geometry and the pool riffle sequence. It has been suggested that meander curvature increases the power expenditure due to secondary currents (Chang, 1984a,b), but such an increase is compensated by an increase in the sediment efficiency associated with transverse bedslope so that uniform utilization of power and continuity of sediment load are maintained through river meanders. Such considerations depend upon energy dissipation rate relationships which are founded upon thermodynamic principles of entropy creation. Davy and Davies (1979) concluded that there is undoubtedly some general principle governing stream geometry and that this principle may be found by further consideration of energy transfer relations in streams. Nevertheless, they believed that the principle of minimum entropy production rate, and the principle that the most probable state of a system is the one which corresponds to maximum entropy are not strictly transferable to stream situations.

A *fourth* approach that can be detected is the way in which stream power varies over time. This is particularly important where human activity is introduced, and in a study of the arroyo systems of the northern part of the Henry Mountains in south central Utah, Graf (1983) has shown that whereas total stream power decreased in the downstream direction during a deposition period which occurred before 1896 when channels were small and meandering, after 1896 total stream power increased in the downstream direction because channels were in the floors of arroyos that confined discharges and resulted in channel erosion and throughput of sediment. He showed that in 1980 the rate of downstream change in total power was intermediate between the depositional conditions of the 1890s and the erosional conditions of 1909 with deposition occurring in the smallest and largest channels but not in the mid basin areas.

Such analyses indicate the potential that an approach based on power has for the investigation of change. Therefore, it is understandable why Thornes (1983) argued that there has been renewed interest in the long-term behaviour of landforms. That is contingent upon a shift from the observation of equilibrium states to the recognition of the existence of multiple stable and unstable equilibria, the bifurcations between them and trajectories connecting them. It will rely upon the adoption of a dynamical approach so that new models of geomorphological evolution can be constructed. However, although the potential for innovative approaches to changes over time may be possible, and may be in line with developments elsewhere in science such as in relation to dissipative structures, such approaches require a sound dynamical foundation upon work and power and the energetics of nature is one with considerable potential. Baker (1983) has speculated that a phenomenal excess of input to a fluvial system could signify that the usual factors do not damp out such a fluctuation and the system enters a high-energy state of disequilibrium. Such a high energy state may lead to chaos and could link with Prigogine's ideas of non-equilibrium thermodynamics.

Not only does an approach based upon power offer considerable potential for investigation of the past but it can also be applied to estimations of future changes and therefore to management. Human activity not only exercises a control upon the power of nature but can also provide additional power sources. Management is often concerned with minimizing power expenditure and suggesting ways in which power expenditure in the natural system can be regulated in a way which is designed to minimize the disruption which can result. Perhaps, therefore, the ways in which power has already been embodied in approaches to the physical environment can provide a basis for the development of new approaches in future years.

1.6 CONCLUSION

It could be argued that an approach to physical geography based upon energetics is not particularly novel because there is evidence that for at least 30 years there has indeed been greater emphasis upon such an approach in the branches of physical geography and this has been reflected in the approach taken in research papers and in books (see Figure 1.1). Perhaps the time is ripe, however, for a statement of a more integrated and unified approach because of the similarities which have been emerging in the branches of physical geography. This is not just another way of focusing upon processes in the physical environment because a more basic approach founded upon energetics or upon the power of nature relates not only to the present but also to past and future environments. Thus in his consideration of the target of space and the arrow of time, Bird (1981) acknowledged that there are areas of geography where mechanistic explanations are valid, and therefore where the laws of physics may be expected to apply. However, physical geography, although requiring a firm scientific foundation as argued by S. Gregory (1978), also needs to be concerned with temporal change and to see the arrow of time not only in relation to the past and to evolution of environment, but also in relation to the future and therefore to prediction. Therefore, if energetics offers a more unified foundation for physical geography this is in keeping with the way in which Medawar (1986) has suggested that sciences are becoming more and not less unified.

However, paradoxically a more unified approach may not necessarily mean that we have less to cover in the scope of physical geography. It may therefore be as Clayton (1985) has suggested that we may have to do less to do anything better and that we must also increase cooperation with other subjects and understand the language that they use. Potentially we may therefore be at the beginning of a very exciting time for physical geography so that the approaches becoming available could help to achieve the goal that F. K. Hare (1985) set as ensuring that the actual future will be better than the actual past. Such an optimistic view was not taken by Haines-Young and Petch (1986) who argue that in physical geography there have been very few advances in our theories about, or our understanding of, the natural world. They concluded that the discipline can boast no major advances.

An alternative view is that what has occurred so far in physical geography has been the foundation for a more integrated phase of development which can now take place. Such an integrated phase could help to resolve the problems of the reductionist approaches that have been adopted until recently; it could avoid the temptation to move towards a very realist approach; and it could present to physical geography a basis for a more organismic approach in which the vital link is the availability of energy, the use of power, the application of energetics and therefore the power of nature. Such a more unified approach is in sympathy with the move towards theories of everything which seek to unify all forces, particles of matter, space, time, and creation into a single descriptive scheme. Pagels (1985) envisages the universe as a giant computing machine with the laws of nature providing the software. It has been suggested by Paul Davies (1985) that such a conception is in tune with the drift of modern science where process is gradually replacing object as the basic concept. In physical geography the power of nature may be the prime focus for process.

Acknowledgements

The great help given by Mr Chris Hill in collecting much of the material that forms the basis for sections of this chapter is gratefully acknowledged.

References

Allen, J. R. L. (1977). Changeable rivers: some aspects of their mechanics and sedimentation. In Gregory, K. J. (ed.) *River Channel Changes*. Wiley: Chichester, 15–45.

Andrews, J. T. (1972). Glacier power, mass balances, velocities and erosion potential. *Zeitschrift für Geomorphologie*, **13**, 1–17.

Andrews, J. T. (1975). *Glacial Systems*, Duxbury Press: Massachusetts, 191 pp.

Atkinson, B. W. (ed.) (1981). *Dynamical Meteorology*. Methuen: London, 228 pp.

Bagnold, R. A. (1960). Sediment discharge and stream power: A preliminary announcement. *US Geological Survey Circular*, 421.

Bagnold, R. A. (1977). Bedload transport by natural rivers. *Water Resources Research*, **13**, 303–312.

Baker, V. R. (1983). Large scale fluvial palaeohydrology. In Gregory, K. J. (ed.) *Background to Palaeohydrology*, Wiley: Chichester, 453–478.

Barsch, D. and Caine, N. (1984). The nature of mountain geomorphology. *Mountain Research and Development*, **4**, 287–296.

Bennett, R. J. and Chorley, R. J. (1978). *Environmental Systems: Philosophy and Control*. Methuen: London.

Bird, J. (1981). The target of space and the arrow of time. *Transactions Institute of British Geographers* NS6, 129–151.

Blong, R. J. (1982). *The Time of Darkness*. University of Washington Press: Seattle and London, 257 pp.

Blong, R. J. (1984). *Volcanic Hazards. A Sourcebook of the Effects of Eruptions*. Academic Press: Sydney, 424 pp.

Bloom, A. L. (1969). *The Surface of the Earth*. Prentice Hall International, London, 152 pp.

Bohm, D. (1980). *Wholeness and the Implicate Order*. Routledge and Kegan Paul: London, 224 pp.

Bormann, F. H. and Likens, G. E. (1981). *Pattern and Process in a Forested Ecosystem*. Springer-Verlag: New York.

Boulton G. S. (1975). Processes and patterns of subglacial sedimentation: a theoretical approach. In Wright, A. E. and Moseley, F. (eds) *Ice Ages Ancient and Modern*. Seel House Press: Liverpool.

Broda, E. (1975). *The Evolution of the Bioenergetic Processes*. Pergamon: Oxford.

Brown, E. H. (1975). The content and relationships of physical geography. *Geographical Journal*, **141**, 35–48.

Budyko, M. I. (1956). *The Heat Balance of the Earth's Surface*. Translated by N. Stephanova, US Weather Bureau: Washington.

Bull, W. B. (1979). Threshold of critical power in streams. *Bulletin Geological Society of America*, **90**, 453–464.

Caine, N. (1976). A uniform measure of subaerial erosion. *Bulletin Geological Society of America*, **87**, 137–140.

Carson, M. A. and Kirkby, M. J. (1972). *Hillslope Form and Process*. Cambridge University Press: Cambridge.

Chang, H. H. (1984a). Minimum stream power and river channel patterns. *Journal of Hydrology*, **41**, 303–327.

Chang, H. H. (1984b). Analysis of river meanders. *Journal of Hydraulic Engineering*, **110**, 37–50.

Cherkauer, D. S. (1978). The effect of urbanisation on kinetic energy distribution in small watersheds. *Journal of Geology*, **86**, 505–515.

Chorley, R. J. (1973). Geography as human ecology. In Chorley, R. J. (ed.) *Directions in Geography*. Methuen: London, 155–169.

Chorley, R. J. (1978). Bases for theory in geomorphology. In Embleton, C., Brunsden, D., and Jones, D. K. C. (eds) *Geomorphology, Present Problems and Future Prospects*, Oxford University Press: Oxford.

Chorley, R. J., Dunn, A. J., and Beckinsale, R. P. (1964). *The History of the Study of Landforms Vol. 1 Geomorphology before Davis*. Methuen: London 678 pp.

Chorley, R. J. and Kennedy, B. A. (1971). *Physical Geography: A Systems Approach*. Prentice Hall: London.

Clayton, K. M. (1985). The state of geography. *Transactions Institute of British Geographers*, NS10 5–16.

Coates, D. R. (1976). Geomorphic engineering. In Coates, D. R. (ed.) *Geomorphology and Engineering*. Dowden, Hutchinson and Ross: Stroudsburg, 3–21.

Coffey, W. (1981). *Geography. Towards a General Spatial Systems Approach*. Methuen: London, 270 pp.

Cooke, R. U. and Warren, A. (1973). *Geomorphology in Deserts*. Batsford: London, 394 pp.

Cruickshank, J. G. (1972). *Soil Geography*. David & Charles: Newton Abbot, 256 pp.

Culling, W. E. H. (1983). Rate process theory in geomorphic soil creep. In Jan de Ploey (ed.) *Rainfall Simulation, Runoff and Soil Erosion*. Catena Suppl., 4, 191–214.

Davidson, D. A. (1978). *Science for Physical Geographers*. Edward Arnold: London, 187 pp.

Davies, G. L. (1968). *The Earth in Decay: A history of British geomorphology 1578–1878*. MacDonald: London, 390 pp.

Davies, J. L. (1980). *Geographical Variation in Coastal Development*, 2nd edn. Longman: London, 212 pp.

Davies, P. (1985). Review of H. Pagels *Perfect Symmetry: The Search for the Beginning of Time*. Times Higher Education Supplement, no. 636, 20 pp.

Davey, B. W. and Davies, T. R. H. (1979). Entropy concepts in fluvial geomorphology: a re-evaluation. *Water Resources Research*, **15**, 103–106.

Douglas, I. (1981). The city as an ecosystem. *Progress in Physical Geography*, **5**, 313–367.

Douglas, I. (1983). *The Urban Environment*. Edward Arnold: London.

Dury, G. H. (1981). *An Introduction to Environmental Systems*. Heinemann: London, 366 pp.

Eddington, A. S. (1958). *The Nature of the Physical World*. Ann Arbor: University of Michigan Press.

Embleton, C. and Thornes, J. B. (1979). *Process in Geomorphology*. Edward Arnold: London.

Erismann, T. H. (1979). Mechanism of large landslides. *Rock Mechanics*, **12**, 15–46.

Fairbridge, R. W. (1980). Thresholds and energy transfer in geomorphology. In Coates, D. R. and Vitek, J. D. (eds) *Thresholds in Geomorphology*. George Allen & Unwin: London.

Ferguson, R. I. (1981). Channel forms and channel changes. In Lewin, J. (ed.) *British Rivers*. George Allen & Unwin: London, 90–125.

Gerrard, A. J. (1981). *Soils and Landforms*. George Allen & Unwin: London, 219 pp.

Gersmehl, P. J. (1976). An alternative biogeography. *Annals Association of American Geographers*, **66**, 223–241.

Graf, W. L. (1979). Mining and channel response. *Annals Association of American Geographers*, **69**, 262–275.

Graf, W. L. (1982). Spatial variation of fluvial processes in semi-arid lands. In Thornes C. E. (ed.) *Space and Time in Geomorphology*. Allen & Unwin: London, 193–217.

Graf, W. L. (1983). Downstream change in stream power in the Henry Mountains, Utah. *Annals Association of American Geographers*, **73**, 373–387.

Gregory, K. J. (1979). Drainage network power. *Water Resources Research*, **15**, 775–777.

Gregory, K. J. (1980). Discussion of drainage network power. *Water Resources Research*, **16**, 1130.

Gregory, K. J. (1982). River power. In Adlam, B. H., Fenn, C. R. and Morris, L. (eds) *Papers in Earth Studies*. Geobooks, 1–20.

Gregory, K. J. (1985a). *The Nature of Physical Geography*. Arnold: London.

Gregory, K. J. (1985b). Environmental management and planning. *Perspectives on a changing geography*. ed. Ashley Kent, Geographical Association, 48–57.

Gregory, S. (1978). The role of physical geography in the curriculum. *Geography*, **63**, 251–264.

Gutenberg, B. (1956). Great earthquakes. *Transactions of the American Geophysical Union*, **37**, 608–614.

Haigh, M. (1985). Geography and general systems theory, philosophical homologies and current practice. *Geoforum*, **16**, 191–203.

Haines-Young, R. and Petch, J. R. (1986). *Physical Geography: Its Nature and Methods*. Harper & Row: London.

Haken, H. (1979). Synergetics and a new approach to bifurcation theory. In Guttinger, W. and Eihemeier, H. (eds) *Structural Stability in Physics*. Springer Verlag: Berlin, 31–39.

Haken, H. (1985). Synergetics—an interdisciplinary approach to phenomena of self organisation. *Geoforum*, **16**, 205–211.

Hare, F. K. (1953). *The Restless Atmosphere*. Hutchinsons Univ. Lib., 192 pp.

Hare, F. K. (1965). Energy exchanges and the general circulation. *Geography*, **50**, 229–241.

Hare, F. K. (1973). Energy-based climatology and its frontier with ecology. In Chorley, R. J. (ed.) *Directions in Geography*. Methuen: London, 171–192.

Hare, F. K. (1966). The concept of climate. *Geography*, 51, 99–110.

Hare, F. K. (1985). Future environments: can they be predicted? *Transactions Institute of British Geographers*, NS10, 131–137.

Harvey, D. (1969). *Explanation in Geography*. Edward Arnold: London.

Hewitt, K. and Hare, F. K. (1973). *Man and Environment: Conceptual Frameworks*. Association of American Geographers, Commission on College Geography, Resource Paper No. 2., 39 pp.

Hobbs, J. E. (1980). *Applied Climatology: A Study of Atmospheric Resources*. Butterworths: London.

Huggett, R. J. (1980). *Systems Analysis in Geography*. Oxford University Press: Oxford.

Ji, S. (1985). The Bhopalator: a molecular model of the living cell based on the concepts of conformons and dissipative structures. *Journal of Theoretical Biology*, **116**, 399–426.

Johnston, R. J. (1979). *Geography and Geographers: Anglo-American Human Geography since 1945*. Arnold: London, 232 pp.

Karcz, I. (1980). Thermodynamic approach to geomorphic thresholds. In Coates, D. R. and Vitek, J. D. (eds) *Thresholds in Geomorphology*. Allen & Unwin: London, 209–226.

Knighton, D. (1980). Drainage network power: Discussion and reply. *Water Resources Research*, **16**, 1129.

Lal, R. (1977). Analysis of factors affecting rainfall erosivity and soil erodibility. In Greenland D. J. and Lal, R. (eds) *Soil Conservation and Management in the Humid Tropics*. Wiley: Chichester.

Leopold, L. B. and Langbein, W. B. (1962). The concept of entropy in landscape evolution. *US Geological Survey Professional Paper*, 500-A, 20 pp.

Lewin, J. (1983). Changes of channel patterns and floodplains. In Gregory, K. J. (ed.) *Background to Palaeohydrology*. Wiley, Chichester, 303–319.

Lindeman, R. L. (1942). The trophic–dynamic aspect of ecology. *Ecology*, **23**(4), 399–418.

Lindsay, R. B. (1976). *Applications of Energy, Nineteenth Century*. Dowden Hutchinson & Ross Inc: Stroudsburg.

Linnhoff, B. (1983). New concepts in thermodynamics for better chemical process design. *Proceedings Royal Society London*, A386, 1–33.

Linton, D. L. (1965). The geography of energy. *Geography*, **50**, 197–228.

Lockwood, J. G. (1979). *Causes of Climate*. Edward Arnold: London, 260 pp.

Marston, R. A. (1982). The geomorphic significance of log steps in forest streams. *Annals Association of American Geographers*, **7**, 99–108.

Mather, J. R., Field, R. T., Kalkstein, L. S., and Willmott, C. J. (1980). Climatology: The challenge for the eighties. *Professional Geographer*, **32**, 285–292.

May, J. P. and Tanner, W. F. (1973). The littoral power gradient and shoreline changes. In Coates, D. R. (ed.) *Coastal Geomorphology*. George Allen & Unwin: London, 43–60.

Medawar, P. (1986). *The Limits of Science*. Oxford University Press: Oxford.

Miller, M. M. (1973). Entropy and self-regulation of glaciers in Arctic and Alpine areas. In Fahey, B. D. and Thompson, R. D. (eds) *Research in Polar and Alpine Geomorphology*. GeoAbstracts Ltd: University of East Anglia, 136–158.

Morris, S. E. (1986). The significance of rainsplash in the surficial debris cascade of the Colorado Front Range Foothills. *Earth Surface Processes and Landforms*. **111**, 11–22.

Odum, E. P. (1968). Energy flow in ecosystems: a historical review. *American Zoologist*, **8**, 11–18.

Odum, H. T. and Odum, E. C. (1976). *Energy Basis for Man and Nature*. Wiley: New York.

Orme, A. R. (1980). Energy–sediment interaction around a groin. *Zeitschrift für Geomorphologie* Suppl.Bd 34, 111–128.

Pagels, H. (1985). *Perfect Symmetry: The Search for the Beginning of Time*. Michael Joseph: London.

Park, C. C. (1980). *Ecology and Environmental Management*. Dawson: Folkestone.

Pethick, J. (1984). *An Introduction to Coastal Geomorphology*. Edward Arnold: London.

Phillipson, J. (1966). *Ecological Energetics*. Institute of Biology. Studies in Biology No. 1 Edward Arnold: London, 57 pp.

Prigogine, I. and Stengers, I. (1984). *Order out of Chaos*. Heinemann: London.

Roberts, J. M. (1980). *The Pelican History of the World*. Harmondsworth: Penguin Books.

Rumney, G. R. (1970). *The Geosystem: Dynamic Integration of Land, Sea And Air*. Wm.C. Brown Company: Dubuque, Iowa, 135 pp.

Runge, E. C. A. (1973). Soil development sequences and energy models. *Soil Science*, **115**, 183–193.

Scheidegger, A. E. (1968). Some implications of statistical mechanics in geomorphology. *Bulletin International Association Scientific Hydrology*, **9**, no. 1, 12–16.

Simmons, I. G. (1978). Physical geography in environmental science. *Geography*, **63**, 314–323.

Simmons, I. G. (1979). *Biogeography: Natural and Cultural*. Edward Arnold: London, 400 pp.

Steinhart, C. E. and Steinhart, J. S. (1974). *Energy: Sources, Use and Role in Human Affairs*. North Scituate, Mass: Duxbury.

Stocking, M. A. and Elwell, H. A. (1976). Rainfall erosivity over Rhodesia. *Transactions Institute of British Geographers*, **1**(2), 231–245.

Stoddart, D. R. (1965). Geography and the ecological approach. The ecosystem as a geographic principle and method. *Geography*, **50**, 242–251.

Strahler, A. N. and Strahler, A. H. (1974). *Introduction to Environmental Science*. Hamilton Pub. Co., Santa Barbara, 633 pp.

Strahler, A. N. (1962). Dynamic basis of geomorphology. *Bulletin Geological Society of America*, **63**, 923–938.

Strahler, A. N. (1980). Systems theory in physical geography. *Physical Geography*, **1**, 1–27.

Strahler, A. N. and Strahler, A. H. (1984). *Elements of Physical Geography*. Wiley: London, 469 pp.

Taube, M. (1985). *Evolution of Matter and Energy on a Cosmic and Planetary Scale*. Springer-Verlag: New York, 280 pp.

Terjung, W. (1976). Climatology for geographers. *Annals Association of American Geographers*, **66**, 199–222.

Thornes, J. B. (1979). Fluvial processes. In Embleton, C. and Thornes, J. B. (eds) *Process in Geomorphology*. Edward Arnold: London.

Thornes, J. B. (1983). Evolutionary geomorphology. *Geography*, **68**, 225–235.

Tinkler, K. J. (1985). *A short history of geomorphology*. Croom Helm: London and Sydney.

Trudgill, S. (1977). *Soil and Vegetation Systems*. Oxford University Press: London.

Weiner, N. (1948). *Cybernetics: or, Control and Communication in the Animal and the Machine*. Wiley: New York, 212 pp.

White, I. D., Mottershead, D. N. and Harrison, S. J. (1984). *Environmental Systems: An Introductory Text*. Allen & Unwin: London.

Williams, P. J. (1982). *The Surface of the Earth: An Introduction to Geotechnical Science*. Longman: London.

Yang, C. T. (1971). On river meanders. *Journal of Hydrology*, **13**, 231–253.

Young, R. W. (1985). Waterfalls: Form and Process. *Zeitschrift für geomorphologie*, Suppl.Bd **55**, 81–95.

Energetics of Physical Environment
Edited by K. J. Gregory
© 1987 John Wiley & Sons Ltd

2

Atmospheric Energetics

B. W. ATKINSON

*Department of Geography and Earth Science,
Queen Mary College, University of London*

2.1 INTRODUCTION

Man has been curious about his natural environment for centuries. The curiosity turned to scientific enquiry at the time of the Renaissance when a myriad of major problems about planet Earth and beyond were recognized. This enquiry became increasingly more specialized over the centuries so that by the 18th century the natural environment as the home of mankind became arguably the primary object of scrutiny. Systematic monitoring of spatial and temporal changes in fauna and flora paralleled a growing awareness that the land itself may well also be subject to changes, albeit on longer timescales. At the same time the vicissitudes of the atmosphere were recognized and indeed, in a very limited number of places, were being instrumentally recorded. But the relative transience and extreme mobility of atmospheric phenomena meant, as they still do, that their investigation was far from easy. On the other hand, accumulated weather lore frequently gave sufficiently accurate forecasts for daily activity in what was still essentially a rural environment. These two factors, no doubt among others, meant that scientific observational study of the atmosphere in the 19th century perhaps lagged behind the physiographical and biological investigation of the solid earth. But from a theoretical viewpoint atmospheric science has had, throughout its history, distinct advantages over its geological, geomorphological and biogeographical counterparts. The mixture of gases that we call air behaves closely in accord with the basic laws of mechanics and thermodynamics. Hence the weather and climate, which largely result from the spatial inequalities of the heat content, mass and motion of air, are themselves explicable in terms of the basic processes.

A major conceptual component of mechanics and thermodynamics is of course energy. Hence in addition to the more conventional aspects of dynamical meteorology that cover relationships between forces and air motion (Atkinson,

1981) the well-established energetic concepts can be applied to atmospheric behaviour. Such applications were initiated in the 19th century and one of the major papers in this field, still relevant today, was due to Margules (1903) in the early years of this century. The increasing emphasis on weather forecasting, largely due to the stimulus of the two world wars, meant that the study of atmospheric energetics remained backstage despite some fundamental contributions (see, for example, Brunt, 1926). It was only in the period after 1945 that resources became available for the fuller study of the less immediately applicable parts of meteorology. In the last four decades a combined theoretical and observational onslaught on the problems of atmospheric energetics has led to enormously increased insight.

This chapter considers both the pure and applied aspects of atmospheric energy. From a pure science viewpoint three topics are treated: the Earth— atmosphere energy budget; the transfers and transformations of energy within the atmosphere; and the spectrum of kinetic energy and its relationship to weather systems in the atmosphere. From an applied science standpoint, the utility of atmospheric energy (sun, wind) and the possible modifications by mankind of the planet's energy budget are considered.

2.2   PURE ASPECTS

Energy is of course the capacity to do work and power is the rate at which work is being done. Remembering that heat is a form of energy, the basic types that we recognize are internal energy (IE), due to the heat content of a body, potential energy (PE), due to the effect of gravity on a body, and kinetic energy (KE), due to the motion of the body. In meteorological terms the body may be a parcel or a column of air or indeed the whole atmosphere. These forms of energy may be variously transferred, transformed and stored. The fundamental problem of atmospheric energetics is to elucidate these changes in the earth–atmosphere system. In what follows a two-tier approach is taken. First, the fairly familiar heat budget of the earth–atmosphere system *in toto* is outlined. Secondly, the transfers (fluxes) and transformations of energy within the atmosphere are considered in more detail.

2.2.1   Energy budgets

Earth–atmosphere and space

The ultimate source of energy for planet Earth is the sun. The energy reaches the Earth by radiation at a rate which varies sufficiently little as still to warrant the name 'solar constant'. The solar constant is defined as the radiant energy flux incident upon a surface perpendicular to the sun's rays, outside the atmosphere, at the average distance between sun and Earth. Relative to an

Table 2.1 Annual mean heat budget of the northern hemisphere (after Atkinson, 1972)

A.	Incoming short-wave solar radiation (W m^{-2})	
1.	Radiation received at upper level of atmosphere	345
2.	Absorption by the atmosphere	
	(a) By ozone (stratosphere)	7
	(b) By water vapour and dust	49
	(c) By clouds	7
	Total absorption	63
3.	Reflection and scattering to space	
	(a) By atmosphere	27
	(b) By clouds	82
	(c) By Earth's surface	13
	Total albedo	122
4.	Absorption by Earth's surface	
	(a) Direct solar radiation	76
	(b) Transmitted by clouds	49
	(c) Scattered	35
	Total absorption by Earth's surface	160
B.	Long-wave terrestrial radiation (W m^{-2})	
1.	Effective radiation of Earth's surface	
	(a) Radiation from surface	397
	(b) Radiation from atmosphere back to Earth's surface	335
	(c) Effective radiative loss from Earth's surface	61
2.	Radiation of atmosphere	
	(a) Earth's surface radiation absorbed by atmosphere	376
	(b) Heat added to atmosphere by convection (84) and conduction (14) from Earth's surface	98
	(c) Short-wave radiation absorbed by atmosphere	63
	(d) Radiation from atmosphere due to heat content: 335 back to Earth (see 1b) 202 out to space	537
3.	Radiation to space	
	(a) Of Earth's surface	21
	(b) Of atmosphere	202
	Total outgoing radiation	223

Note: Total outgoing long-wave radiation (223) + total reflected short-wave radiation (122) = total input of short-wave radiation (345) imaginary disc of equal diameter to the Earth this flux is about 1380 W m^{-2}. Relative to the surface of the Earth (four times the area of the disc) the flux is about 345 W m^{-2}. It is this flux of heat energy which is available to sustain planet Earth. It takes but little reflection, however, to realize that if such a flux continued with no compensation, the planet would soon become totally inhospitable to life as we know it. This would happen in a few weeks. In fact, of course, the planet loses heat at about the same rate as it gains it, thus giving the notion of a heat (energy) balance. Fortunately the atmospheric temperatures resulting from this balance are quite acceptable to mankind. It is highly probable that the solar constant is in fact significantly variable with the consequence that Earth's global climate may range from conditions slightly warmer than present day to glacial regimes with mean atmospheric temperatures some 15–20°C lower than now.

Upon reaching the atmosphere the heat energy is absorbed, reflected and scattered to varying degrees by different components of the Earth–atmosphere system. Table 2.1 shows the disposition of these effects. Of the incident radiation only about 18% is absorbed by the atmosphere, principally by water vapour. Some 35% is reflected back to space by both the atmosphere and Earth's surface. This is known as the planetary albedo. Hence only 46% of of the incident short-wave solar radiation is absorbed by the Earth's surface. This low value of surface receipt and the lack of absorbtivity of the atmosphere are perhaps the two most striking facets of the input of radiant energy to the Earth–atmosphere system.

Turning to the output side of the balance, the mean surface temperature of the Earth (287 K) means that it radiates in long wavelengths at a rate of 397 W m^{-2}. Because the atmosphere is an efficient absorber of long-wave radiation only 21 W m^{-2} of the Earth-surface radiative flux escape to space; 376 W m^{-2} are absorbed. The atmosphere also gains heat from convection and conduction (non-radiative heat transfers) from the Earth's surface to add to that from the direct solar flux (63 W m^{-2}). Thus the atmosphere radiates at a rate of 537 W m^{-2}, 335 W m^{-2} returning to the Earth's surface and 202 W m^{-2} being directed to space. The total outward flux of heat is listed in the last part of Table 2.1 and it is evident that a balance is achieved.

Within the Earth–atmosphere system

In addition to the global-scale budget outlined above, a further, and no less important, heat balance exists *within* the Earth–atmosphere system. The sphericity, rotation and angle of ecliptic of the Earth result in significant spatial and temporal variations in receipt of solar radiation even at the top of the atmosphere. The atmosphere itself further modifies these variations by absorption, scattering and reflection. It was shown above that both the Earth and atmosphere radiate in long wavelengths. The Earth's surface absorbs a flux of 160 W m^{-2} and effectively loses only 61 W m^{-2}. Consequently the Earth's

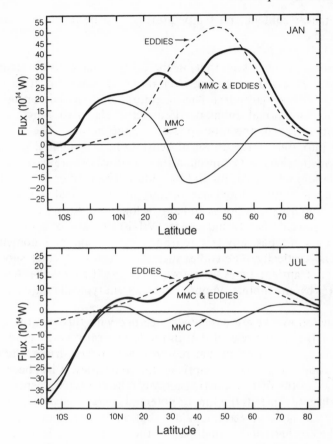

Figure 2.1 Fluxes of total energy (potential energy plus sensible heat plus latent heat) due to mean meridional (MMC) and eddy circulation for January and July. The fluxes are integrated over the layer 1012.5 to 75 mb. Northward fluxes are positive. Units, 10^{14} W (After Oort and Rasmussen, 1970)

surface receives a positive net radiative flux of about $100\,\mathrm{W\,m^{-2}}$ and this surplus is spread over nearly all latitudes. In contrast the atmosphere absorbs $63\,\mathrm{W\,m^{-2}}$ of short-wave radiative flux but loses about $162\,\mathrm{W\,m^{-2}}$ of long-wave flux (discounting net transfer by conduction and convection) thus having a deficit of about $100\,\mathrm{W\,m^{-2}}$ which is spread uniformly over the latitudinal range. This means that the Earth's surface must lose heat to the atmosphere and simple addition of the atmospheric and earth fluxes gives the latitudinal variation of the net radiation budget within the whole Earth–atmosphere system. This distribution is the most profound one in meteorology because it generates the whole general atmospheric circulation. Without such a circulation and its transfers and transformations of energy, the latitudinal gradients of heat content and temperature would be much greater than are presently found.

Energy fluxes   We have seen that, in the Earth–atmosphere system, there is a pronounced excess of incoming over outgoing radiation at low latitudes and a deficit at high latitudes. The magnitude and spatial configuration of the mean net radiational imbalance is directly related to a strong poleward transport of energy in the atmosphere and the oceans. The transport can be divided into four parts: those of sensible heat, potential energy, latent heat, and kinetic energy. The mechanisms of transport are basically twofold: by mean meridional circulation and by eddies. The former is essentially what happens with a simple, closed overturning in the vertical in the meridional plane. The latter are the cyclones and anticyclones that form such an important part of extratropical weather and climate. Whilst they are, of course, important weather bearers they basically exist to transfer heat, water and momentum both poleward and vertically. The eddies themselves can be further subdivided in standing and transient types, the former exemplified by the Azores high, and the latter by any mobile synoptic-scale system typically crossing the north Atlantic.

In all that follows we are basically concerned with fluxes in the meridional direction for the simple reason that in the long term it is the gradient of the net radiation in this direction that requires, and indeed drives, the transfers. Within the total energy flux, comprising the four types, it has been shown by Starr (1951) that the flux of kinetic energy can be neglected. Hence in the data below 'total energy' means the sum of potential energy, sensible heat and latent heat. Figure 2.1 shows the latitudinal distribution of the meridional flux of total energy in the northern and a small part of the southern hemisphere in January and July. It is clear that winter is the season of greatest transfer with maximum amounts of about 40×10^{14} W at latitudes 50 to 60°N. Note also that all the flux is toward the north. In July the flux in the southern hemisphere part of the figure is towards the south and the poleward flux in the northern hemisphere has less latitudinal variability with a general level of about 12×10^{14} W. Figure 2.1 also reveals the contribution of eddy and mean meridional fluxes. It is quite clear that in the winter hemisphere the eddies are the primary mode of poleward transfers particularly poleward of about 20°N. In contrast the mean meridional circulation is the major mechanism south of 20°N, albeit transferring only about half the energy transferred by eddies. Between latitudes of about 30 to 60° the mean meridional circulation transfers energy equatorward, hence going some way to offset the poleward eddy transport. In July the mean meridional circulation is very weak and the greater part of the transfer is achieved by the eddies. The January pattern dominates the annual regime and leads to the widely recognized view that mean meridional transfers are of primary importance in the tropics and eddies are paramount in the extratropics. This notion was

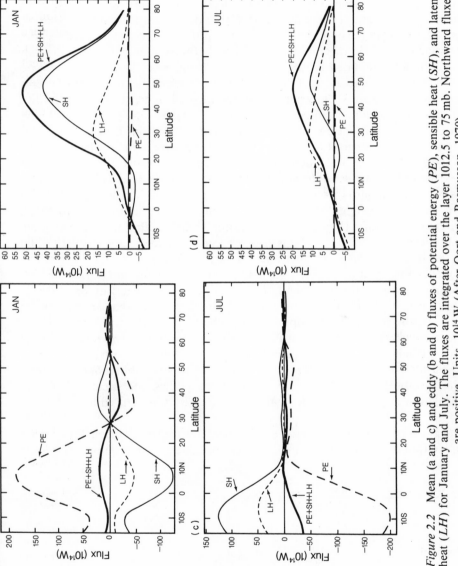

Figure 2.2 Mean (a and c) and eddy (b and d) fluxes of potential energy (*PE*), sensible heat (*SH*), and latent heat (*LH*) for January and July. The fluxes are integrated over the layer 1012.5 to 75 mb. Northward fluxes are positive. Units, 10^{14} W (After Oort and Rasmusson, 1970)

suggested from theoretical work by Jeffreys (1922) over 60 years ago and has been confirmed by the observational studies of the past three decades.

The total values in Figure 2.1, informative as they are, conceal further interesting results when the separate forms of energy are analysed. Figure 2.2 shows that the total mean meridional flux in January is northward only because the flux of potential energy in that direction dominates the fluxes in sensible and latent heat in the opposite direction. The process is the same, with directions reversed in July. Further, we see that the eddy fluxes in both January and July are strongly dominated by the sensible and latent heat fluxes.

The vertical distributions of the eddy fluxes of each of the four types of energy are shown in Figure 3. Only annual figures are used to save space. The vertical variation of mean meridional fluxes are not shown as values at one level have little meaning in isolation. They are but local contributions from a given layer to the vertically integrated flux. It is clear from Figure 2.3 that the largest values of the eddy fluxes of both potential and kinetic energy are an order of magnitude less than those of sensible and latent heat. Sensible heat (Figure 2.3a) is transferred poleward throughout the depth of the troposphere north of 20° with two maxima at heights of 850 mb and 200 mb between latitudes 50 to 60°N. Latent heat transfer (Figure 2.3b) has a simpler distribution with its maximum at low levels around 30°N. This is largely because of two factors: first, the water comes from the Earth's surface and would therefore be expected to give high content values near the surface; secondly, both the standing and transient eddies are particularly active between latitudes 30 and 60°N. The flux of potential energy is quite different from the others. Figure 2.3c shows that weak poleward flux in the middle-extratropical atmosphere is overlain by maxima of equatorward flux at about 200 mb in the higher latitudes. The flux of kinetic energy (Figure 2.3d) reverts to a more conventional pattern with the maximum at 200 mb over latitude 30°N probably reflecting jet streams in mobile cyclones. The transport of kinetic energy by standing eddies and mean meridional circulation is generally much smaller than that by transient eddies.

Energy transformations   We have seen earlier how the spatial inequality of input and output of heat energy in the Earth–atmosphere system requires and results

Table 2.2   Energy content of atmosphere

Type	Amount per unit area (J m^{-2})	Amount per unit mass (J kg^{-1})	Total amount (TJ)	Per cent of total
Potential	567.5×10^6	0.55×10^5	2.89×10^{11}	25.28
Internal	1674.8×10^6	1.61×10^5	8.54×10^{11}	74.67
Kinetic	1153.4×10^3	1.11×10^2	5.88×10^8	0.05
	22434.5×10^5	2.16×10^5	1.14×10^{12}	100.00

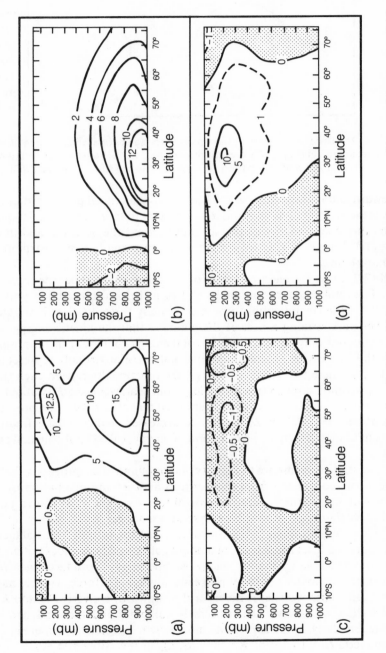

Figure 2.3 Mean annual northward transfer by all eddies of (a) sensible heat; (b) latent heat; (c) potential energy; and (d) kinetic energy. Units, J kg^{-1} m s^{-1} (After Oort and Rasmusson, 1971)

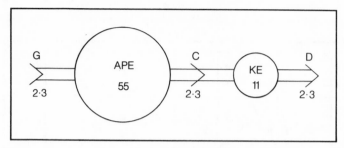

Figure 2.4   The basic energy cycle of the atmosphere. Values of available potential energy (*APE*) and kinetic energy (*KE*) are in units of 10^5 J m^{-2}. Values of generation (G), conversion (C) and dissipation (D) are in W m^{-2} (After Lorenz, 1967)

in poleward and upward transfers of heat. But, as a central part of atmospheric science is the analysis of air motion, it is important to investigate the relationships between the kinetic energy of this air motion, and the potential and internal energies resulting primarily from the temperature and heating distribution in the atmosphere.

Table 2.2 shows the energy content of the atmosphere under the three basic forms. The total content is about 10^{12} TJ or a (British) quadrillion joules, of which 25% is in the potential form and virtually three-quarters is in internal form. The very striking aspect of these data is that the kinetic energy of the winds represents only 1/2000 th of the total energy content of the atmosphere. Clearly the sum of the potential and internal energies virtually represents the total. In fact these two forms of energy are functionally related such that in a dry, hydrostatic atmosphere extending from sea-level the *PE* is about two-fifths of the *IE*. Table 2.2 suggests that in the real atmosphere the proportion may be about 35%. *PE* and *IE* increase or decrease together and it is frequently convenient to regard them as a single form of energy, called total potential energy (*TPE*).

The essential problems are to identify and to quantify the processes that lead to generation, conversion from one form to another and dissipation of these energies within the atmosphere. In general terms we know, as shown earlier, that the Earth–ocean–atmosphere system exchanges total energy with its environment only through radiation. Because of the overall heat balance the system gains and loses roughly equal amounts of *IE* and hence does not undergo any net long-term change in total energy. Kinetic energy is produced (generated) or destroyed (dissipated) only by processes involving a force. Hence, motion with or against gravity, that is downward or upward, converts *PE* into *KE* or *KE* into *PE* respectively. The process is adiabatic and thermodynamically reversible. Similarly, in the horizontal, motion with or against a pressure gradient force converts *IE* into *KE* or *KE* into *IE*, again reversibly and adiabatically. Motion against the force of friction, resulting in frictional heating, also converts

KE into *IE*. Three decades ago Lorenz (1955) reappraised the way in which *TPE* was converted into *KE*, and introduced the notion of available potential energy (*APE*) and unavailable potential energy (*UPE*). The *APE* of the atmosphere in any given state is the difference between *TPE* in the given state and the value it would have after an adiabatic redistribution of mass had produced a horizontal stratification. Lorenz showed that only a very small fraction of the *TPE* was available for conversion to *KE*. In fact *APE* represents about 1% of *PE* and 0.25% of *TPE*.

The introduction of this powerful notion, along with earlier estimates of energy dissipation rates by friction, led to the relatively simple concept of the basic energy cycle of the atmosphere as shown in Figure 2.4. This figure has two major components: the amounts of *APE* and *KE*; and the rates of energy generation (*G*), conversion (*C*) and dissipation (*D*). It is clear that the amount of *APE* is five times that of *KE*. This large discrepancy between what is available and what is 'used' is largely due to the *APE*, as defined, being an upper bound on the amount of energy available for conversion to *KE* and perhaps also to some storage effect. The equivalence of the generation, conversion and dissipation rates simply affirms the necessity for and reality of long-term balance in the transfers and transformations that are the very stuff of mechanisms in the natural environment. Temporal and spatial variations of *G*, *C* and *D* occur within the atmosphere as is illustrated below. The magnitude of each of *G*, *C* and *D* is a rough measure of the efficiency of the atmosphere. We showed earlier that the energy flux into the atmosphere was about 350 W m^{-2}. With an albedo of about 120 W m^{-2} this leaves only 230 W m^{-2} for use by the planet. Figure 2.4 shows that only 1% of that flux is required to generate the *APE* which, by conversion at the same rate to *KE*, balances the frictional dissipation rate of the latter. In view of the windspeeds that can be experienced on this planet, perhaps we should be grateful that the atmospheric engine runs at this very low rate.

Valuable as Figure 2.4 is, it does not reveal the mechanisms of *G*, *C* and *D*. The *APE* is generated by heating of the air resulting in differences between *TPE* and *UPE*, manifest in differences in the potential temperature fields of the given and reference states. In turn, there are two main methods by which the heating can produce the *APE*: by heating warmer regions and cooling cooler regions at the same height, thereby increasing horizontal temperature contrasts; and by heating lower levels and cooling upper levels, thereby decreasing the static stability. The conversion of *APE* to *KE* occurs when cold air sinks and warm air rises at the same level. In such a situation the centre of gravity of the air falls, hence the *PE* decreases and the *KE* increases (Margules, 1903). The dissipation of *KE* is largely by friction at the Earth's surface.

Further examination of these processes has led to the concepts of zonal and eddy energy, particularly as applied to *APE* and *KE*. Zonal *APE* (*ZAPE*) is the amount of *APE* that would exist if the field were replaced by its zonal

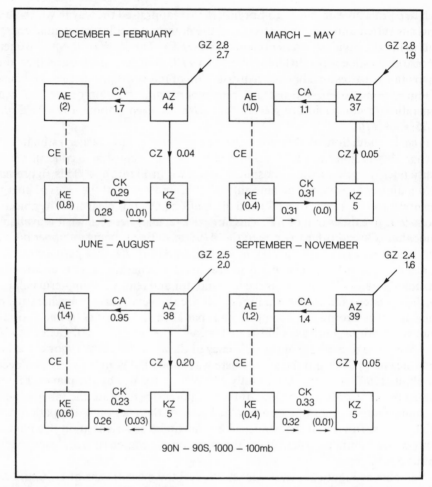

Figure 2.5 The energy cycle for four 3-month periods, 90°N–90°S, 1000–100 mb. Units: Contents in 10^5 J m^{-2}; conversions in W m^{-2}. Parentheses indicate quantity available only from 90°N–30°S. Contributions to CK are divided into transient eddy (lower left) and standing eddy (lower right). Key to letters: AE: eddy available potential energy; AZ: zonal available potential energy; KE: eddy kinetic energy; KZ: zonal kinetic energy; CA: conversion of zonal available potential energy into eddy kinetic energy; CK: conversion of eddy kinetic energy into zonal kinetic energy; CZ: conversion between zonal available potential and kinetic energies; GZ: generation of zonal energy (After Newell *et al.*, 1969)

(latitudinal) average and eddy *APE* (*EAPE*) is the excess of *APE* over *ZAPE*. Similarly zonal *KE* (*ZKE*) is the amount of *KE* that would exist if the existing zonal averaged motion but no eddy motion were present, and eddy *KE* (*EKE*) is the excess of *KE* over *ZKE*. This further subdivision of energy into zonal and eddy components goes some way to revealing the types of weather patterns that are primarily responsible for the conversion of *APE* to *KE*. The two most obvious possible processes are a general sinking of cold air in high latitudes and rising of warm air in low latitudes (revealed by zonal data), and a sinking of cold air in the colder parts of cyclones and anticyclones and a rising of warm air in the warmer portions at the same latitudes (revealed by eddy data).

Figure 2.5 shows the amounts of zonal and eddy *APE* and *KE*, and their generation, conversion and dissipation for four three-month seasons. The major differences between the seasons are the magnitudes of the variables and the change in the direction of *CZ* in March to May. Otherwise the major theme is that *ZAPE* is generated (*GZ*) by the broad latitudinal gradient in heating. Eddy transfer of heat toward colder latitudes means that *ZAPE* is converted (*CA*) into *EAPE*. In turn *EAPE* is converted (*CE*) into *EKE*, largely by extratropical cyclones and anticyclones. Whereas Newell *et al.* (1969) refrained from computing this conversion rate for want of good information on vertical velocities, Oort (1964) suggested a figure of 2.2 W m^{-2} for the year as a whole. The conversion from *ZAPE* to *ZKE*, by vertical overturnings in the Hadley and Ferrel cells, is very small and may indeed reverse its direction in one season. Perhaps the most intriguing conversion is from *EKE* to *ZKE* which suggests that large-scale eddying motion is an un-mixing process — a manifestation of what has been called 'negative viscosity' (Starr, 1968). Figure 2.5 includes no dissipation rates, but Newell *et al.* (1969) gave figures ranging from 4.1 to 10.4 W m^{-2} recognizing that they are larger than would perhaps be expected. Oort (1964) gave a dissipation of *EKE* of 1.8 W m^{-2} and of *ZKE* of 0.5 W m^{-2}. Before leaving this subject it is worth noting that Newell *et al.*'s values for *EKE* (in the range of 0.5×10^5 to 1.0×10^5 J m^{-2}) are substantially less than Oort's (1964) figure of 7×10^5 J m^{-2}. Newell *et al.* gave no explanation of this but agreed that, depending upon the method of calculation, it is possible that *EKE* would be approximately equal to *ZKE* with a value of about 5×10^5 J m^{-2}. Oort's figure for *ZKE* was 8×10^5 J m^{-2}.

2.2.2 Energy spectrum

Although the kinetic energy of the atmosphere is but a very small fraction of the total energy content, the motions that are its essence also form the core of meteorology. It has long been recognized that insight into weather and climate depends upon an understanding of how and why air moves as it does within the atmosphere (Atkinson, 1981). A major stepping-stone towards such understanding is an appreciation of scale, which can be approached from both theoretical and empirical standpoints. A major empirical tool is the variance spectrum. This allows the variance of a temporal or spatial series to be apportioned according to frequency or wavelength. When the variance is that of wind speed it becomes a statement about kinetic energy. Hence, several tens of velocity spectra constructed over the past three decades allow an appreciation

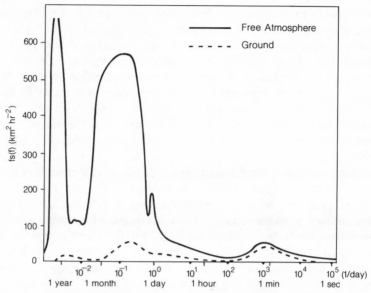

Figure 2.6   Spectra of horizontal, zonal wind and hence of kinetic energy. Spectra in the free atmosphere (solid line) and near the ground (dashed line) (After Vinnichenko, 1970)

of the distribution of atmospheric kinetic energy according to either frequency or wave length. All the spectra show large maxima of energy at wavelengths of several thousand kilometres and periods of 1 day to months. In increasingly smaller and ephemeral systems the energy content falls off markedly. Recent results (see, for examples, van Zandt, 1982) suggest that the fall is continuous, whereas earlier spectra (van der Hoven 1957; Vinnichenko, 1970) suggested a local maximum at periods of a few minutes (Figure 2.6). The so-called 'spectral gap' between the two maxima in Figure 2.6 understandably led to a three-tier hierarchy of atmospheric systems: synoptic-scale, meso-scale and micro-scale. Although the gap is losing its credibility as better spectra are being produced, the shape of the graphs is still highly suggestive to those seeking underlying mechanisms. For example, van Zandt (1982) claims that spectra in the range 2 min to 1 day (1–1000 km roughly equivalent wavelength) show remarkably consistent slopes which in turn are very suggestive of gravity (buoyancy) waves as major features and transfer mechanisms in the atmosphere at this particular scale.

2.3   APPLIED ASPECTS

2.3.1   Power from the atmosphere

In recent years the world has become more conscious of the increasing need for and cost of energy for human activities. At a fundamental level energy is required of course for all human activity but at a more practical level we are concerned primarily with energy sources that give heat, light, mobility and other so-called necessities of modern life. In this practical domain the atmosphere has two major sources that may be tapped: winds and solar radiation.

Wind power

We have seen earlier that the winds represent a vast amount of kinetic energy in the atmosphere — of the order of 10^{20} J. The power of the wind is calculated as follows. A mass of air, m, moving with velocity, v, has kinetic energy $\frac{1}{2}mv^2$. The mass of air meeting an area, A, normal to the wind, in one second is ϱAv where ϱ is the density of air. Hence the kinetic energy of air meeting the area A in 1 s, (that is, its power) is $\frac{1}{2}\varrho Av^3$. So a uniform wind of 10 ms^{-1} meeting an area of 10 m^2 would generate 6 kW. Noting that the power of the wind varies with the cube of the wind speed, if we assume a mean density of 0.5 kg m^{-3} and a mean wind speed of 20 ms^{-1}, the power of the horizontal winds of the atmosphere is about 2 kW m^{-2} or about 2.4×10^{15} W for an atmosphere of 30 km depth. As we can extract power to any significant practical degree from winds near the surface (say 10 m), then only 0.3×10^{-3} of the total power is available. This gives a figure of about 10^{12} W, in agreement with Hewson's (1975) estimate.

Although the efficiency of windmills (or aerogenerators) is still not known as well as we would like, it is probable of the order of 30–40%. Hence, any calculation and analysis of the distribution of wind power is essentially that of the available, not extractable power. Within the United Kingdom the wind power in the Southern Uplands of Scotland has been estimated at an annual average of 174 W m^{-2}, ranging from 319 W m^{-2} in December to 77 W m^{-2} in July (Duncan, 1977). The figures are comparable to, but slightly less than those found in the United States, where the annual average ranged from over 400 W m^{-2} in Wyoming to less than 50 W m^{-2} in the south-east of the country (Reed, 1974). All these figures were derived from wind data recorded at, or standardized, to a height of 10 m. In fact, wind speeds increase with height, roughly in accord with a 1/7 power of height ($z^{1/7}$). Hence the power of the wind, being a function of the cube of its speed, increases with $z^{3/7}$. Clearly the higher the generator the better. Unfortunately this, of course, usually means a more expensive structure.

The full assessment of the real practicability of extracting wind power consistently in significant amounts remains incomplete. Of the many examples in the United States, it has been estimated that about 100 large 80 m diameter units in a line 4 km long would be necessary to generate sufficient electricity for the annual demands of Albuquerque, New Mexico (population at time of estimate, 322 000) (Reed, 1974). In the United Kingdom Duncan (1977) estimated that a windmill with a radius of 3 m mounted at a height of 10 m would be necessary to produce the 420 W of electricity required in July in southern Scotland. Such a device would produce a surplus of electricity, but barely enough for space heating in winter. A major problem is the storage of the electrical power generated by wind. Consequently, wind power is likely to remain for some time a complementary source to augment the major source of fossil fuel. On a small scale, rural areas isolated from national supplies, are the most likely places to benefit from the application of small, wind-driven electricity generators.

Solar power

Of the actual and possible uses of solar radiation—many of the actual uses having been outlined earlier in this chapter—the possibility that it can be used directly in mankind's activities for heating/cooling is particularly attractive. In the higher latitudes, where the sun's power is less than in lower latitudes augmentation of space and water heating could be a boon. Whereas in lower latitudes, the high temperatures, largely due to high solar income, may be offset to some extent by air conditioning, itself partially powered by the very same radiation.

The average amount of solar radiation reaching the Earth's surface is 160 W m^{-2}. The variations around that mean are well exemplified by the 95 W m^{-2} or so received in the uplands in the United Kingdom, and this figure itself conceals seasonal variations. In winter space and water heating for domestic purposes can be as high as 1680 W. To provide such an amount of power could require horizontal solar heat-collecting panels (with an efficiency of 50%) of an area of 160 m^2. If the panels were inclined so as to be normal to the sun's rays the mean annual total solar radiation is increased by about 60%. In addition, if the panels are fixed, they should face south in the northern hemisphere or north in the southern hemisphere. Duncan (1977) showed that a 70 m^2 heat collector, at an angle of 35°, would supply 1680 W for heating of a domestic building in the Scottish Southern Uplands in December.

2.3.2   Man's impact on atmospheric energetics

For the greater part of his history, man has been at the mercy of the elements, not least the atmosphere. Even today, we are far from independent of nature's vicissitudes and wherever possible attempt to extract benefits, such as outlined in the previous section. In recent decades, however, mankind's technologies have appeared to have deleterious effects upon nature to the extent that serious assessment of these effects has been high on the international scientific agenda. In terms of the energetics of the atmosphere four effects have received closest attention: gaseous (particularly carbon dioxide, CO_2) pollutants; particulate pollutants; thermal pollution; and modifications of the Earth's surface. All of these affect, directly or indirectly, the magnitudes of the terms in Table 2.1.

Carbon dioxide is a small (0.03% by volume) constituent of the atmosphere, but despite this it plays an important role in determining the temperature of the planet. As shown in Section 2.2.1 (p.34), it absorbs and emits long-wave radiation and helps the atmosphere to act as a blanket around the Earth. Such evidence as exists shows that CO_2 increased over the decade 1958–69 at about 0.2% per year or about 0.7 ppm out of 320 ppm. In this period about half of the input remained in the atmosphere. If such a situation continued it would lead to an increase of CO_2 over present levels of about 20% in the year 2000. Faced with such figures the immediate qualitative conclusion is that atmospheric temperatures would increase. A major way of acquiring quantitative estimates is to run mathematical models of the general atmospheric circulation with different CO_2 contents. Several experiments of this kind have been done over that past decade and the results are far from clear-cut (Liss and Crane, 1983). Most experiments involve the extreme case of doubling the CO_2 content of the atmosphere and from such it appears that the global surface temperature might increase by 1.5 to 3.0°C with values as high as 7°C in the lower, high-latitude atmosphere (Schneider, 1975).

Particulates influence atmospheric energetics through their interaction with both solar and terrestrial radiation. Clearly natural processes (such as volcanoes) have injected material into the atmosphere throughout time but mankind is now significantly adding to these amounts, particularly locally, such as in urban areas. In short-wavelength solar radiation, particle layers in the atmosphere may change both the global albedo and the absorption of radiation by the atmosphere. Using reasonable assumptions on the properties of particles it appears that an increase in particle load increases the global albedo and so must cool the planet. Extreme actual and possible examples are severe volcanic eruptions (such as Mount St Helens in the United States) and the so-called 'nuclear winter'. In these cases it is still not yet possible to be certain either that temperatures did fall significantly after Mount St Helens, or that a winter would ensue from a large-scale nuclear war. Research into the effects of particulates continues apace at present.

Thermal pollution, or the emission of waste heat, is probably the least potentially dangerous of man's direct injections into the atmosphere. Compared to the solar radiation at the Earth's surface (about $160\,W\,m^{-2}$), mankind's energy production distributed evenly over the whole globe is about $15 \times 10^{-3}\,W\,m^{-2}$. Locally (such as in Los Angeles) the waste heat approaches 5% of the incident solar radiation, and it is at this scale that such emission may become a problem in the next century.

Perhaps mankind's most potent influences on the atmosphere lie in his capacity to change the nature of the Earth's surface. The properties that affect energetics are: reflectivity (albedo), heat capacity and conductivity, availability of water, availability of dust, aerodynamic roughness and emissivity in the infrared wavelengths. Changes in any of these properties will affect the disposition of energy, in a few cases increasing losses, in most of the others, changing the distribution and use of the energy. Few, if any, of these effects have been estimated in a serious manner.

References

Atkinson, B. W. (1972). Physical climatology. In Bowen, D. Q. (ed.) *A Concise Physical Geography*, Hulton Educational Pub. Ltd: Amersham, 4–32.

Atkinson, B. W. (1981). Weather, meteorology, physics, mathematics. In Atkinson, B. W. (ed.) *Dynamical Meteorology: An Introductory Selection*, Methuen: London, 1–7.

Brunt, D. (1926). Energy of the earth's atmosphere. *Philosophical Magazine* 7, 523–532.

Duncan, C. N. (1977). Solar and wind power — some meteorological aspects. *Weather*, 32, 451–456.

Hewson, E. W. (1975). Generation of power from the wind. *Bulletin of the American Meteorological Society*, 56, 660–675.

Jeffreys, H. (1922). On the dynamics of winds. *Quarterly Journal of the Royal Meteorological Society*, 48, 29–46.

Liss, P. S. and Crane, A. J. (1983). *Man-made Carbon Dioxide and Climatic Change. A Review of Scientific Problems*. Geobooks: Norwich.

Lorenz, E. N. (1955). Available potential energy and the maintenance of the general circulation. *Tellus*, 7, 157–167.

Lorenz, E. N. (1967). *The Nature and Theory of the General Circulation of the Atmosphere*. Tech. Note No 218, TP 115, World Meteorological Organization: Geneva.

Margules, M. (1903). Über die Energie der Sturme, *Jahrb Zentralanst*, Vienna, 1–26.

Newell, R. E., Vincent, D. G., Dopplick, T. G., Ferruzza, D., and Kidson, J. W. (1969). The energy balance of the global atmosphere. In Corby, G. A. (ed.) *The Global Circulation of the Atmosphere*. Royal Meteorological Society: London, 42–90.

Oort, A. H. (1964). On estimates of the atmospheric energy cycle. *Monthly Weather Review*, 92, 483–493.

Oort, A. H. and Rasmusson, E. M. (1970). On the annual variation of the monthly mean meridional circulation. *Monthly Weather Review*, 98, 423–442.

Oort, A. H. and Rasmusson, E. M. (1971). *Atmospheric Circulation Statistics*. NOAA Professional Paper 5, US Department of Commerce, National Oceanic and Atmospheric Administration, Rockville.

Reed, J. W. (1974). Wind power climatology. *Weatherwise*, 27, 236–242.

Schneider, S. H. (1975). On the carbon dioxide — climate confusion. *Journal of the Atmospheric Sciences*, 32, 2060–2066.

Starr, V. P. (1951). Applications of energy principles to the general circulation. In Malone, T. F. (ed.) *Compendium of Meteorology*, American Meteorological Society: Boston, 568–574.

Starr, V. P. (1968). *Physics of Negative Viscosity Phenomena*. McGraw-Hill Book Co.: New York.

van der Hoven, I. (1957). Power spectrum of horizontal wind speed in the frequency range from 0.0007 to 900 cycles per hour. *Journal of Meteorology*, 14, 160–164.

van Zandt, T. E. (1982). A universal spectrum of buoyancy waves in the atmosphere. *Geophysical Research Letters*, 9, 575–578.

Vinnichenko, N. K. (1970). The kinetic energy spectrum in the free atmosphere — 1 second to 5 years. *Tellus*, 22, 158–166.

Energetics of Physical Environment
Edited by K. J. Gregory
©1987 John Wiley & Sons Ltd

3

Erosional Energetics

C. JANE BRANDT AND JOHN B. THORNES

Department of Geography, University of Bristol

3.1 INTRODUCTION

Hillslope forms are produced as a result of the application of solar radiant energy, relief potential energy and precipitation kinetic energy and their interaction with the physical and chemical properties of the surface, including the vegetative cover. The efficiency of application, defined as work done relative to energy applied, is reflected in the amount of form change in a given amount of time or, alternatively, the amount of sediment produced in the drainage basin. We shall see that the efficiency is generally low, reflecting the loss of energy in overcoming friction in various forms. Weathering energetics are discussed by Ross in Chapter 5 (p. 123), and the global efficiency of erosional processes as reflected by sediment transfers in rivers by Walling in Chapter 4 (p. 97). This chapter is mainly concerned with the role of energy and its dissipation at the hillslope scale and over relatively short periods of time. Rather than adopt a budget approach, which would be premature given the very limited data available, we consider the nature of the balance of energy application and conversion in terms of the processes involved. The chapter will not be concerned with the production of relief energy through orogenic and epeirogenic uplift. The most up-to-date review (Chorley, Schumm, and Sugden, 1985) reaffirms the view, held since the time of W. M. Davis, that uplift is localized, sporadic and relatively fast, when compared to the rates of denudation. Moreover, hillslope forms are more likely to reflect channel activity and hillslope processes and their energetics rather than the differential rates of uplift (Ahnert, 1970).

Geomorphologists and those concerned with soil erosion have so far paid little attention to the energy aspects of erosional processes. This is because it is widely presumed and asserted that the efficiency of energy application (the amount of work done relative to the amount of energy applied) is very low and there is some evidence to support this view, at least as far as hillslopes are concerned (see below, p. 75). Coupled with this is the difficulty in obtaining

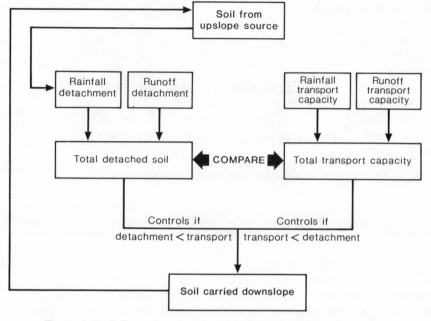

Figure 3.1   Soil erosion model of Meyer and Wischmeier (1969)

realistic data, since the magnitude of observational errors are quite large when compared with, say meteorology or climatology. Therefore, even if it were relevant, it will probably not be possible to build up an erosional energy budget for a long time to come. It is for these reasons that the emphasis is laid on the processes of energy transformation on hillslopes rather than its partition.

Soil erosion has been studied for a long time and there are at present two main approaches to the problem. The first is pragmatic and designed for application. It consists of identifying the main factors affecting soil loss and presenting them in an empirical equation, the universal soil loss equation, the terms of which are parameterized from plot data collected from around the world but especially from the United States. This approach was introduced by Wischmeier and Smith (1958) and recently reviewed by Mitchell and Bubenzer (1980). The second approach, the rational approach, is to build up a model of soil erosion by applying physical laws to the component processes which can be tested in the field as independent elements or as interacting processes. It is this approach, as formulated by Meyer and Wischmeier (1969) which provides a suitable vehicle for the discussion of erosional energetics.

The basic model, schematized in Figure 3.1, originated in the ideas of Ellison (1944) and was formalized by Meyer and Wischmeier (1969). In this model the main processes are detachment and transport by rainfall and runoff and these processes are interrelated. Runoff is divided into rill and inter-rill flow. Rills are ephemeral channels which remain on the surface for only one or two seasons and are then broken down by frost heave, weathering or agricultural activity or filled in by splash transport from the adjacent inter-rill areas. In the Meyer and Wischmeier model the detachment capacity is compared with the transporting capacity and the lesser of the two is taken to be the amount of erosion at a point. In a given slope section, the amount of material in transport increases by the amount of material detached provided that there is spare transporting capacity or that extra transporting capacity is generated in that slope section, for example, by more power becoming available. Foster and Meyer (1972) improved this approach by making the transport in a section increase according to the magnitude or the 'spare' capacity, which is defined as the difference between the actual transporting capacity and the amount of sediment being transported. This formulation modifies the old idea of erosion being limited by the availability of material (weathering or detachment limited) or limited by the transporting capacity (transport limited) which can be traced at least as far back as the works of G. K. Gilbert.

The model has also been modified by appreciation that splash and wash processes strongly interfere with each other to provide intermediate processes called splash-creep and wash-creep. It has also been realized that although the processes can be theoretically defined and separated in the laboratory they are almost impossible to separate in a field context. Moreover, the introduction of vegetation substantially changes both the energy available and the nature of some of the processes. Despite these limitations, a modified version of the Meyer and Wischmeier (1969) model, as shown in Figure 3.2, provides an appropriate framework for the consideration of erosional energetics.

The final point to be made by way of introduction is that the management of erosion largely consists of managing the energetics of the system. This is normally achieved by changing the vegetation cover, reducing the runoff volume (and hence mass) and velocity, or adjusting the relief or slope to change the

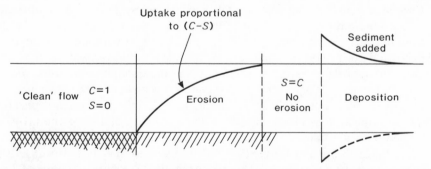

Figure 3.2 Foster and Meyer (1972). Modification of the Meyer and Wischmeier Model. In this modification, the rate of uptake by entrainment is made proportional to the difference between sediment in transport and the stream capacity

potential and kinetic energy. Some of these control mechanisms are considered in Section 3.5.

3.2   RAINFALL EROSIVITY

The *erosivity* of rainfall may be defined as its ability to erode the soil surface. This is distinguished from *erodibility*, which is the propensity of soil to erosion. Many indices of erosivity have been proposed, some a composite of other rainfall properties such as intensity (Ellison, 1944; Ekern, 1950, 1953; Kneale, 1982). However, if all erosional processes are to be assessed together it is important to have a common index so that the different processes may be compared. Riezebos and Epema (1985) have compared the correlation between measurements of splash detachment and transport, and rainfall kinetic energy and momentum. They found that generally the kinetic energy had the higher correlation.

3.2.1   Rainfall kinetic energy

The kinetic energy of a rain storm (E, joules) may be calculated from the sum of the energy of all raindrops (*e*) such that:

$$E = \Sigma e_n \tag{1}$$

where *n* is the diameter of the raindrop and e_n may be calculated from:

$$e_n = \tfrac{1}{2} m v^2 \tag{2}$$

where m(g) is the mass of the drop and v(m s^{-1}) its velocity. To use this equation, several simplifying assumptions have to be made about the shape and fall velocity of the drops. One necessary assumption is that drops maintain a constant shape during fall. However, single drop studies have shown that as raindrops greater than 0.34 mm diameter fall, their shape oscillates from an oblate to a prolate spheroid (Pruppacher and Pitter, 1971). According to Ekern (1953), maximum transport coincides with drops which impact with a minimum area. Riezebos and Epema (1985), moreover, showed that if an individual drop strikes the ground with a prolate shape, as opposed to an oblate shape, the amount of soil detached is two to three times greater. However, compensation for the changing drop shape has so far only included calculations using a spherical or hemispherical shape on impact. It is also assumed that the velocity of fall of the drops is dependent on their height of fall alone (Chapman, 1948; Dohrenwend, 1977; Mosley, 1982). Gunn and Kinzer (1949) demonstrated that the majority of the drop sizes occurring naturally in rain reach terminal velocity after falls of 8 m. During rain storms however there may be up or down draughts of air which change the velocity of the drops. Hitherto this problem has remained intractable.

3.2.2 Drop size distributions

Any storm is made up of a number of drops of different size, and the range of sizes may vary throughout the storm, depending in part on the type of storm, larger drops tending to occur in convective storms (Horton, 1948). The maximum drop size is determined by the physical stability of the falling drop and has been estimated for still air to be between 8.6 mm (Pruppacher and Pitter, 1971) and 13.1 mm (Klett, 1971). However, drops of this size are not usually encountered in storms where collision and coalescence of the drops produce, for a given rainfall rate, raindrop spectra of remarkable similarity (Srivastava, 1971).

The actual drop-size distribution for any storm may be obtained either from direct measurement or from established links between rainfall intensity and rainfall drop-size distribution. Although the former may be more desirable in terms of an accurate calculation of the value for kinetic energy, the latter has more potential for use in cases where detailed data is not available.

The earliest work on the frequency distribution of raindrop sizes has been reviewed by Mason (1957). Marshall and Palmer (1948) used the data of Laws and Parsons (1943) to determine the number of drops for any given size at different intensities. They found that except for small sizes the drop-size distribution can be represented by:

$$N(x) = N_0 e^{-bx} \qquad (3)$$

where $N(x)$ is the number of drops per unit volume having a diameter between x and $x + dx$ (cm), $b = 4I^{-0.21}$ where I is rainfall precipitation (mm hour^{-1}) and N_0 is a constant 0.08. It appears from these formulae that the number of drops rapidly increases as the drop size decreases at a rate depending on the intensity of the rain. Since Marshall and Palmer, other workers (for example, Mason and Ramandham, 1953; Houze *et al.*, 1979) have found that despite overestimating the numbers of smaller drops, the Marshall–Palmer distribution fits their data for different rain types well.

In contrast to work relating intensity to the frequency of drop sizes, Laws and Parsons (1943) and Best (1950a) considered drops in terms of their volume. They all related volume to intensity because for many purposes the number of drops of a particular size is considered less important than the volume of water comprising drops of that size (Figure 3.3). Best (1950a) derived the following equations:

$$F(x) = 1 - \exp\left[-(x/a)^n\right] \qquad (4)$$
$$a = AI^p \qquad (5)$$

where $F(x)$ is the fraction of liquid in the air comprised of drops with a diameter of less than x (mm), I is the rainfall intensity (mm hour^{-1}) and n, A and p

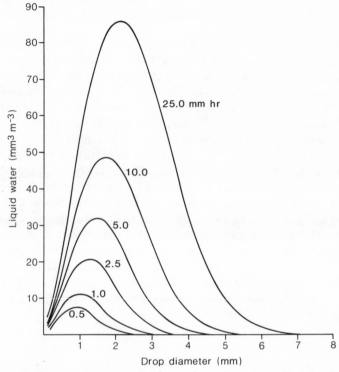

Figure 3.3 Distribution of water (mm³ mm⁻¹) against drop diameter for different rainfall intensities (from Best, 1950a) see also Figure 4.2 (p.92)

are constants, taken for all the data then available on drop-size distributions to be 2.25, 1.3 and 0.232 respectively.

In contrast, and where such experimentation is feasible, the drop-size distribution changes throughout a storm may be measured directly. Although time-consuming this has the advantage of monitoring changes in drop-size distribution throughout a storm more accurately and is the only method for obtaining drop-size distributions which have been altered by a canopy.

3.2.3 Measuring storm energy

There are many methods for determining the sizes of raindrops. These include scanning clouds with radar (Mason and Andrews, 1960), measuring the impact of a drop on a pressure-sensitive membrane (Schindelhaur, 1925 (reported in Mason, 1957); Palmer, 1963) and the use of a photoelectric spectrometer (Mason and Ramandham, 1953). These techniques all involve a difficult calibration and the most successful techniques are those which make a record of each drop. The early development of techniques for measuring the drop-size distribution of rainfall were described by Laws and Parsons (1943) who favoured the use of the flour-pellet method, a technique more recently used by Best (1950b) and Kneale (1982). Flour is sieved into a pan and then exposed to rain for a known time. The raindrops form pellets in the flour which may be separated by sieving after they have been baked in an oven. It is then possible to relate the pellet size to the original drop size by measuring the size of pellets formed by drops of known diameter.

Early users of the paper staining technique are quoted in Laws and Parsons (1943). More recent users include Mosley (1982), Noble and Morgan (1983) and Brandt (1986). The technique involves colouring absorbent paper with a dye which reacts with the raindrops to leave a circular stain. As in the flour-pellet method, the size of the stain may be related to the original drop size by considering the sizes of stains produced by drops of known diameter.

Repeated sampling of a storm by any of these techniques will define the change in drop-size distribution throughout the storm. If discrete sampling is used, in order to calculate the kinetic energy of a storm, the drop-size distributions in the unsampled gaps must be obtained from the samples, using an interpolation technique (Brandt, 1986). The relationship between kinetic energy and intensity for each sample may also be used to establish more general relationships (Mosley, 1982).

As pointed out earlier, the majority of workers to date have assumed that raindrops fall at their terminal velocity on impact with the ground. The velocity of falling drops has been measured experimentally and derived from formulae that predict the terminal velocity.

Empirical data has commonly been used to develop the formulae for predicting drop size. Best (1950b) used unpublished data by Davies (1942), and Schottman (1978) used the data of Laws (1941). Small drops with a diameter of up to 0.05 mm can be assumed to fall according to Stoke's law (Best, 1950b). However, the changes in shape of falling drops of diameter greater than 1 mm make formulation more difficult, and therefore the resultant calculations for terminal velocity depend on the shape the drops are supposed to assume. While Best (1950b) persevered with spherical drops, Schottman (1978) assumed they were of hemispherical shape.

Early work measuring the terminal velocity of water drops was carried out by Lenard (1904) and Flower (1928). However, Hall (1970) has since shown that their values differ by as much as 15 per cent from values calculated later. Laws measured the velocities of water drops with diameters between 1.25 and 6 mm, falling in still air from heights of 0.5–20 m. Where the heights of fall are less than 8 m, if there is a tree or plant canopy, the velocity of fall may be calculated from this height. Probably the most extensive and accurate set of measurements have been made by Gunn and Kinzer (1949) who measured the velocity of fall of a range of drops after a fall height of 20 m.

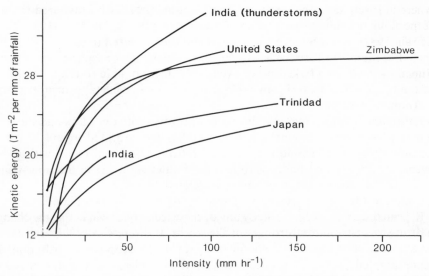

Figure 3.4   Kinetic energy of rainfall against rainfall intensity (from Jansson, 1982)

A large body of work has concentrated on the relationship between kinetic energy and rainfall intensity. This avoids the need to calculate the kinetic energy of rainfall from first principles. Similarly, the relationship between throughfall energy and intensity has been calculated for a number of plant canopies. Chapman (1948) measured the drop-size distribution of rainfall. Combining the drop diameters with Laws's (1941) data, he calculated the kinetic energy for each intensity of rain falling at up to 150 mm/h. Working empirically, Wischmeier and Smith (1958) found a very high correlation between kinetic energy of a storm, or part of a storm, and the rainfall intensity of the form

$$E\,(\text{mm}) = a + b\log I \qquad (6)$$

where $E\,(\text{mm})$ is the energy due to each mm of rainfall $(\text{J mm}^{-1}\,\text{m}^2)$ and I is the rainfall intensity (mm h^{-1}). Morris (1986) found values of a and b to be 11.9 and 8.7 respectively, and Brandt (1986) for tropical storms in Amazonas found values of 15.5 and 2.46. Kinnel (1973) used drop-size data from five rain types to ascertain a general relationship between kinetic energy and rain intensity and reported a correlation coefficient between the two of 0.996. He later (1981) combined data collected in Zimbabwe and Miami concerning the energy/unit quantity of rain (J_{ra}) and the intensity of rain (I) and found a relationship of the form

$$J_{\text{ra}} = z(1 - p.e^{nI}) \qquad (7)$$

where z, p and n are empirical constants with values of 22.2, 0.894 and 0.477 respectively for the Zimbabwe case. Figure 3.4 shows the energy/unit depth of rain rising with an inverse exponential rate for several different locations, including Kinnel's data for Zimbabwe. By contrast Mosley (1982) reported a linear relationship between kinetic energy and intensity such that $J(\text{J min}^{-1}\text{m}^{-2} = 19.19 + 0.461$ (cm/h^{-1}).

To obtain the kinetic energy of a storm, the kinetic energy for each distinctive intensity increment is obtained by multiplying the kinetic energy by the rainfall amount for that increment. The total energy of a storm is then obtained by accumulating the kinetic energy for each distinctive intensity increment of the event.

3.2.4   The effects of vegetation

The drop-size distribution of rain and its consequent erosive power may be changed by a vegetation canopy. Chapman (1948) reported that there was a tendency for a more or less equal distribution by volume of rain among drops of all sizes. Thus under the canopy a much larger portion of the total volume of water falls in the form of large drops. Ovington (1954) measured drop sizes in open rain and under saturated canopies of 13 tree types and found that in the forest plots the majority of the falling water drops were within the range of drops falling in the open, but that there were also an additional number of larger drops which constituted the greatest weight of water.

In contrast to rainfall, in which the kinetic energy varies with the rainfall intensity, the kinetic energy of throughfall from a plant canopy tends to remain constant (Chapman, 1948; Tsukamoto, 1966 (reported in Dohrenwend, 1977); Mosley, 1982; Brandt, 1986). The canopy changes the energy by changing the drop-size distribution of the rainfall. Since large drops lose proportionally less kinetic energy to air resistance, the shattering of a large drop to many smaller drops of the same total volume will decrease the kinetic energy of the rain. Conversely the coalescence of a number of larger drops will increase the kinetic energy for the same volume of water.

Typical values of kinetic energy for rainfall are about 15–24 J mm^{-1} m^{-2} although for short periods during tropical storms the values may exceed 50 J mm^{-1} m^{-2} (Brandt, 1986). The kinetic energy of throughfall from vegetation canopies, which remains constant with changing rainfall intensity, tends to be consistently higher than the kinetic energy of rainfall. Chapman (1948) estimated that only at rainfall intensities of 50 mm h^{-1} would the kinetic energy of rainfall exceed that of throughfall from a red pine canopy. Mosley (1982) found in a New Zealand beech forest that the estimated kinetic energy of throughfall was 28.6 J mm^{-1} m^{-2} for a 51 mm, 36 h storm, as compared with a value of 18.82 J mm^{-1} m^{-2} above the canopy. For a deciduous oak canopy and frontal rain, Brandt (1986) found average values calculated from 71 samples of raindrops taken throughout a number of storms, of 12.02 J mm^{-1} m^{-2} and 14.97 J mm^{-1} m^{-2} for rainfall and throughfall respectively. Under tropical rain forest canopy, and for dry season storms, rainfall values of 14.75 J mm^{-1} m^{-2} were increased by a single-layered, high forest canopy to 27.02 J mm^{-1} m^{-2} and a full-thickness, low rain forest canopy to 19.96 J mm^{-1} m^{-2}.

Soil splash may be considered as an indicator of the amount of kinetic energy applied to the ground surface and measurements of soil splash in controlled conditions under forest canopies have confirmed this change in erosivity of rainfall by a canopy. Considering a high canopy, Mosley (1982) measured both drop-size distribution and surface splash from sand-filled cups. He found that the kinetic energy of the throughfall was considered 1.5 times higher than that of the rainfall in the open, and that mean values of splash were 3.1 times greater under the canopy than in the open. Wiersum (1985) also measured the erosive power of falling water drops using splash cups filled with sand under an *Acacia* plantation in Java. He measured an increase of 157 per cent in erosive power/unit precipitation falling on the splash cups. Brandt (1986) measured sand splash under a single-layered canopy of more than 8 m high in tropical rainforest near Manaus and found that splash was increased in one case to 660 per cent of splash by rainfall.

By contrast and because of the effect of the height of the canopy, a low vegetation cover tends to reduce the kinetic energy of the rain and thereby protects soil from splash. Hudson (1971) reported on work contrasting erosion on a bare soil surface with a surface protected by a fine-mesh wire gauze representing a vegetation cover of 100 per cent. The soil loss from the bare plots was more than 100 times that from the protected plot. Reductions in splash detachment under low-growing crops compared with bare ground have been measured by Sreenivas, Johnston, and Hill (1947), Bollinne (1978), and Morgan (1982). Complementing these works, Quinn and Laflen (1981) reported that maize with a canopy cover of 36–78 per cent reduced the kinetic energy by 38 per cent to 66 per cent. McGregor and Mutchler (1978) found reductions of 75–90 per cent. Combining the two approaches, Noble and Morgan (1983) measured both kinetic energy and soil splash under Brussells sprouts. The plants were found to reduce the storm energy of rainfall at the ground surface to a mean of 34 per cent. Detachment of soil under the plant ranged from 0.91 to 1.54 kg m^2 compared with a value of 1.21 kg m^2 when no plant was present. In effect, the reduction in kinetic energy did not in this case result in a reduction in the mean amount of splash. They ascribe this to the effects of the drips swamping the splash cups causing soil to be washed instead of splashed out.

However, it would seem that the presence and degree of cover of a litter layer, being the last barrier before the soil, must ultimately govern the amount of kinetic energy the rain possesses when it reaches the soil. At present much of the work in this area has been speculative. Chapman (1948) stated that 'the protection afforded the mineral soil under forest stands . . . arises not from

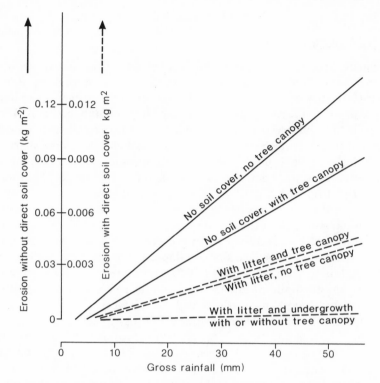

Figure 3.5 Soil erosion (kg m⁻²) versus rainfall (based on Wiersum, 1985, for a variety of overlying vegetation layers)

the overhead canopy breaking the impact of the raindrops, but rather from the presence of a layer or layers of unincorporated organic matter on the mineral soil surface'. Tsukamoto (1966) concluded his work stating that the controlling role of the forest vegetation on erosion rates is due 'mainly to the litter layer on the surface of the forest soil'. Dohrenwend (1977) concluded that the extremely resilient litter layer absorbs virtually all the kinetic energy of the impacting raindrops.

Wiersum (1985) determined experimentally the loss of soil from forest plots both with litter and with the litter removed. The *Acacia* stand decreased the amount of water reaching the soil by 11.8 per cent but increased the erosive power by 24.2 per cent. The litter layer caused erosion to decrease by as much as 93.5 per cent in comparison with erosion on a bare soil plot. The presence of undergrowth decreased erosion by a further 3.7 per cent. Figure 3.5 shows the rates of increase in erosion with increase in rainfall for different vegetation covers. Wiersum concluded that the direct soil cover was the single most important vegetation factor in protecting the soil.

3.3   RAINFALL EROSION

3.3.1   The splash process

The appearance of a water drop splash has been described in two ways. First, from observations of high speed photographs, and second by numerical simulation. Harlow and Shannon (1967) used the numerical 'marker and cell' technique in which the moving drop is divided into a number of cells. For each cell the velocity and pressure are calculated for some small time t, and then recalculated for $t + 1$. Repeated calculations reveal changes in the shape of the drop. Mutchler (1967) used a high speed photographic technique. Both approaches reveal similar patterns of movement (see Figure 3.6). Following the impact of the drop with a hard surface, horizontal flow away from the point of impact is resisted by the surface which causes the splash sheet to rise and form a corona. As the sheet flows and becomes thinner, fluid threads are formed which in turn break into droplets. Once the energy of the water drop is expended, sheet flow ceases and the splash shape collapses. When the drop falls into deep water Park, Mitchell, and Bubenzer (1982) report a return flow on the collapse of the corona to the point of impact. Such a return flow, called a Rayleigh jet, probably does not exist on relatively dry soil surfaces.

The same two methods, numerical and photographic, have been used to estimate the location and size of the forces involved during a splash. Ghadiri and Payne (1979), using the high speed photographic technique to calculate the velocity of the water, measured the impact of water drops on several different surface textures. On all surfaces and within 0.1 ms of impact, a sheet of water moved outwards at 45° with a velocity of three times that of the impact velocity. The impact stress, force per unit perimeter, and kinetic energies calculated are presented in Table 3.1. It was originally assumed that the impact stress was uniformly distributed over the surface of the impact. However, the increase in impact stress with a decrease in drop size suggested that the impact stress was higher round the periphery of the drops.

Huang, Bradford, and Cushman (1982) used the 'marker and cell' technique in calculating the change in velocity and pressure in different components of

Table 3.1   Impact stress, force per unit of perimeter and kinetic energy of falling drops of different size (Ghadiri and Payne, 1979)

Drop diameter (mm)	Stress (kN/m^2)	Force/perimeter (N/m)	Kinetic energy (mJ)
6.1	15	23	1.3
5.4	25	33	1.5
5.3	26	39	1.5
4.6	27	31	1.1
4.5	25	28	0.9
3.3	31	25	0.4

the drop on impact. Like Ghadiri and Payne they found that as the impact progressed, velocities of components at the contacts were laterally dominant, but with values ranging from near zero at the contact circumference. After $18\,\mu s$ a jet stream started to develop with a velocity twice the impact velocity. Within $1\,\mu s$, extremely high pressures were calculated at the impact surface. The stress distribution was not uniform on the impact surface, the maximum stress being calculated at the contact circumference. Huang, Bradford, and Cushman (1982)

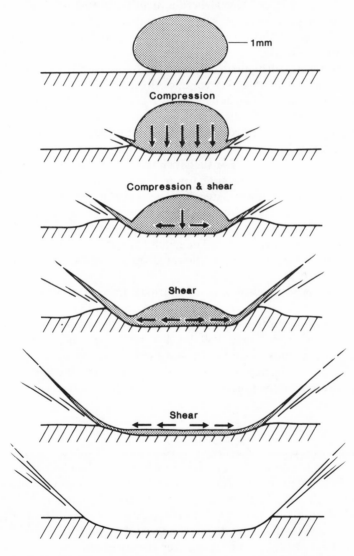

Figure 3.6   Schematic diagram of a rainsplash (from Al-Durrah and Bradford, 1982)

believed that it is the large shearing stress of the lateral jet moving across the irregular soil surface that is the most important process in soil detachment by raindrops.

Using high speed photography Mihara (1951) noted that at early stages of impact a drop simultaneously penetrates the sand surface and spreads out. The depth of penetration depends on the sand surface conditions, especially the water content. The bottom surface of the crater was found to be convex rather than flat. The crater depth decreased with increasing sand compaction and decreasing velocity of the impacting drop. The cavity diameter was slightly larger than the diameter of the drop and increased as the velocity of the drop increased.

From the results of high speed photography of splash on soils of different shear strength and from soil mechanics principles, Al-Durrah and Bradford (1982) proposed a mechanism of splash due to raindrops impacting onto saturated soil surfaces. At the instance of impact the pressure and shear stress distribution are symmetrical about the centre of impact. As already shown above, the peak pressure occurs at the circumference of the contact surface and diminishes in about 6–10 μs. For such high rates of load application on saturated soils there will not be enough time for drainage to take place since the external loads change at a rate much faster than the rate at which the pore pressure can dissipate. Under this condition the impact area will be strained vertically. This change in shape will be compensated for by the development of a bulge around the perimeter of the depression. The magnitude of the vertical strain and area of application are determined by the raindrop size and velocity at the instance of impact.

The compressive stresses are transformed into shear stresses due to the lateral jetting water with a greater velocity than the impact velocity. The shear stresses act on the bottom and sides of the cavity and on the circular bulge. The amount of detachment will be determined by the magnitude of soil deformation that took place earlier and by the cohesive forces resisting shear stress. The greater the depth of cavity and size of bulge, the larger is the splash angle as a result of the greater interception of lateral flows.

3.3.2   The impact on erosion

The impact of the rainfall energy on erosion depends on the balance between the erosivity of the rain and the erodibility of the soil. Bubenzer and Jones (1971) calculated the kinetic energy applied by artificial rainfall on the soil surface and found a correlation coefficient of 0.84 between the energy and weight of soil moved. Ellison (1944) developed an empirical expression for rainsplash as a function of fall velocity, drop diameter and rainfall intensity. Later, in analysing this data, Park, Mitchell, and Bubenzer (1982) found that the ratio of mass of splashed soil to mass of raindrops increased non-linearly with impact velocity.

The magnitude of the vertical strain from the raindrop and the area of impact are determined by the raindrop size and velocity.

Opposing the impacting forces are the strength properties of the soil. Al-Durrah and Bradford (1982) found that soil strength influenced the angle of the splashed drops. The greater the depth of water penetration into the soil surface, the larger the volume of soil pushed to the bulge around the perimeter, and hence the greater the splash angle. The size of the splash angle was shown to be related to the weight of material splashed with higher angles associated with more soil splash. Splash weight was also related to the process of detachment due to lateral flow across the crater boundary. It may be concluded from their work that the lower the shear strength of the soil, the higher the splash angle and the greater the amount of splash.

Apart from the influence of crater size on splash angle and hence splash weight, there is also a relationship between factors affecting the growth of the corona and splash weight. Ghadiri and Payne (1979) reported that the time during which splash droplet formation continues is important in determining the size of droplets released. If splash ceases before droplets are formed, which are equal in size to the surface solid particles, then no movement of the solid occurs. On fine sand (200 μm) at 10 cm water tension, splash ceased 3 ms after the impact. The largest drops formed were 0.2 mm diameter. On the same surface covered with 2 mm of water, splash continued for 80 ms and droplets of 2 mm diameter were formed. The largest amounts of solid particles were lifted when the surface was just saturated but not submerged, conditions which combine moderate duration of splash with no cushioning effect of surface water (Park, Mitchell, and Bubenzer, 1982).

These results support the work of many earlier researchers (for example, Palmer, 1965) who reported increases in amounts of splash with slight increases of water level. Park, Mitchell and Bubenzer (1982) ascribe this change in ability to move material to a change in the method of movement or erosion domain. From the drop–solid interaction where momentum is conserved like colliding billiard balls, to the drop–liquid–solid interaction where the hydrodynamic effects of the lateral jets are included.

3.3.3 The efficiency of rainfall energetics

An expression for the mechanical efficiency of the rainsplash process may be obtained from measurement of energy used in detachment and transport and in the kinetic energy applied. Caine (1976) expressed the amount of energy used by the change in potential energy (ΔPE) of the particles such that:

$$\Delta PE = mg. \; \Delta h \tag{8}$$

where m is the mass of the eroded material (kg) and g is the acceleration due to gravity. *h* is given by $L.\sin\theta$, with L the downslope distance moved and θ the slope angle. Hence the efficiency of the rainsplash process may be defined as

$$E(rs) = \Delta PE/RE \qquad (9)$$

where RE is the rainfall energy. Using such expressions the efficiency of the rainsplash processes has been calculated at 0.2 per cent (Pearce, 1976), between 0.06 per cent and 0.13 per cent (Morgan, 1978) and 0.05 per cent (Morris, 1986).

A remarkably small amount of the kinetic energy applied by rainfall to the soil surface is therefore utilized in splash movement. The reasons for this are implicit in the description of the splash process given above. A force operating perpendicularly on a body, causing motion in the opposite direction to the force will be less efficient than one which operates in the same direction in doing work, much of the energy being transformed to frictional heat losses during impact. Energy is transformed in compacting the soil surface and in forming the splash crater. After impact the lateral jets encounter frictional resistance at the crater surface. As indicated earlier a soil with low cohesion absorbs less energy on compaction and offers less resistance to the lateral shear and hence the amount of splash is higher.

Despite the low efficiency of rainsplash processes, Morris (1986) finds that in an event where the cohesion of surface material is low and vegetation sparse, rainsplash may account for all the surficial debris movement.

3.4   RUNOFF ENERGETICS

3.4.1   The energy of flowing water

Once rainfall and throughfall accumulates in sufficient quantity to overcome detention storage it starts to flow downhill, and the potential energy available to it by virtue of its height is transformed into kinetic energy. In this part of the chapter we consider first the energy of sediment-free water and how it is utilized in overcoming the various resistances to flow under different types of flow condition. Because of its relevance to hydraulic and aeronautical engineering, this branch of science has a huge and diverse literature, only a minute part of which can be considered here. It is also one of the most difficult areas of science because it involves one of the great unsolved problems of our time, the dynamics of turbulent fluids. Although turbulence is a common experience it is mathematically rather difficult to handle. When this is coupled with the fact that water in nature is nearly always carrying sediment, some simplification is necessary if progress is to be made. This takes the form of simple boundary conditions (channels are assumed to have simple forms), simple water conditions (water is usually considered to be uniform and steady or at least only gradually varied), and simple sediment conditions (particles are usually regarded as of a single size and non-cohesive).

The energy of moving water at any location is made up of potential energy due to elevation, kinetic energy related to the velocity of the fluid and its pressure. In pipes, the pressure can be greater than atmospheric and the flow is called pressure flow. In open channels the only pressure force is that due to the depth of the flowing water and the flow is called gravitational. This sum is given by the equation:

$$E = \text{elevation energy} + \text{pressure energy} + \text{kinetic energy}$$

$$= Wz + Wh + (\tfrac{1}{2})\,(W/g)v^2 \tag{10}$$

whose terms are all measured in joules. It is more usual however to express the equation in linear terms as energy head in which case the individual terms also have to be written as head, by dividing through by W (whose units are newtons) giving the units as joules/newtons = metres $(J/N = m)$. This results in the expression:

$$E = z + d + v^2/2g$$

in which the pressure term has been replaced by the water depth relative to the bottom of the flow (to allow for gravitational flow with a free water surface) and the gradient of flow is assumed to be small. When the energy is related to the bed of the channel as in gravitational flow it is usually referred to as the specific energy. This relationship is shown graphically in Figure 3.7 (a). The expression for the conservation of energy in fluids is derived from Bernoulli's theorem, which says that in an ideal fluid, with no energy losses and no viscosity, the energy head at two points in a stream is given by:

$$v_1^2/2g + d_1 + z_1 = v_2^2/2g + d_2 + z_2 \tag{11}$$

In practice, there is always the transformation of energy to other forms not covered by the ideal case, and the difference in the energy head at two places enables us to quantify these losses. Thus for a real fluid, the equation has to be modified to give:

$$(z_1 + d_1 + v_1^2/2g) - (z_2 + d_2 + v_2^2/2g) = \text{loss } (Hf) \tag{12}$$

where Hf is the energy transformation (sometimes called loss) to these other forms. This is shown diagrammatically in Figure 3.7(b), and as head, again has the unit as metres. In this diagram it is assumed that the energy loss occurs at a uniform rate between the two points. The dotted line representing the

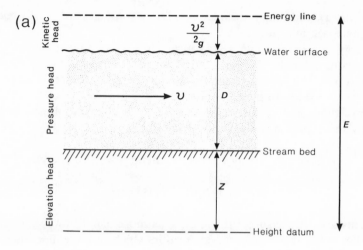

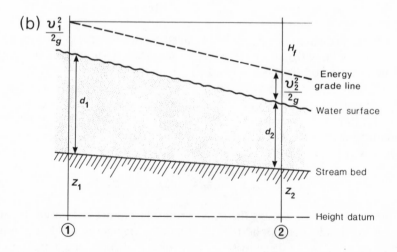

Figure 3.7 (a) Terms in Bernoulli formulation of energy components (b) Diagram illustrating formulation of head loss (Hf) between two points in a uniform steady flow

differences in energy head between points A and B is called the energy grade line or energy slope or friction slope, and like the water surface and the bed slope of the stream it is dimensionless. By observing the terms of equation (12), suitably modified to take account of channel shape, we can calculate head (and hence energy losses) between two points.

If we consider the energy at a particular cross-section and how it varies with depth, z is constant and the energy head is given by

$$E = d + v^2/2g = d + \frac{Q^2}{2gA_2} \tag{13}$$

where, in the second expression we have replaced v by Q/A with A the area of the cross-section. The equation can be solved for v or Q and shows that for a given Q there are two values of d which satisfy the equation except at point A (Figure 3.8) where one value of D corresponds to one value of v, Q and of E. This depth is called the critical depth, and the corresponding velocity the critical velocity. At depths above this value the velocity is less than the critical velocity and the flow is said to be subcritical. Flow at the lower stage is said to be supercritical. At the critical depth, specific energy is at a minimum. If the equation is differentiated and set to zero (to find the minimum E) this occurs where $v\sqrt{gd} = 1$. Since $v\sqrt{gd}$ is the definition of the Froude number, the critical depth occurs at a Froude number of unity. This occurs rarely in rivers but can be induced in man-made structures. Its importance lies in the fact that energy expenditure is rather different in supercritical when compared with subcritical flow.

This set of relationships is very important in engineering practice and has been found to be robust across a wide variety of flow conditions. A very readable outline of their application to hydraulics is given in Kanen (1986).

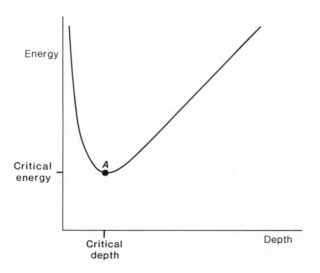

Figure 3.8   Curve of specific energy according to depth at a particular location

3.4.2   Energy losses

The transformation of stream energy into heat and noise takes place by a variety of mechanisms. Of these, most attention has been paid to what is called skin resistance because of its importance in aircraft design and in artificial channels. This is the drag which the boundary imposes on the fluid, and vice versa. When water is flowing at low velocities it can be thought of as like a series of layers sliding over one another, the streamlines operating in parallel (the flow is uniform) and at constant velocity (the flow is steady). With an increase in velocity, exchange of fluid mass and momentum occurs between the layers and the flow is described as turbulent. A measure of the condition of the flow is given by the Reynolds number (Re) which is the ratio of inertial to viscous forces. This amounts, in water flow, to the ratio $\varrho u R/v$, where ϱ is the density of the water, u its velocity and v the viscosity of the water; R is the hydraulic radius. For small speeds, small depths and very low densities, the Reynolds numbers are very small, the viscous effects are most important and inertial forces are negligible. Typically in open channel flow, the flow is laminar up to Re about 2000, transitional from 2000–10 000 and turbulent above that value, whereas on hillslopes the Reynolds numbers are very small and the flow is, by definition, laminar (but see discussion below). In turbulent flow the inertial component (proportional to u^2) is more significant.

To overcome the drag on the boundary the stream has to expend energy and in uniform steady flow this accounts for that part of the value of Hf which appears in the Bernoulli equation. The energy is transferred from the free surface flow down through the boundary layer where the flow velocity is progressively

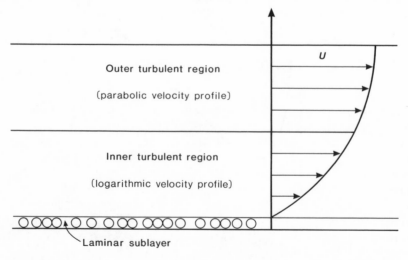

Figure 3.9   Velocity profile in a turbulent stream with a rough boundary

diminishing (Figure 3.9). This transfer or momentum flux is identified by the shear stress within the flow and it is proportional to the momentum gradient, which in turn is a function of the velocity gradient. In other words, the greatest transfer occurs where the velocity gradient is steepest, and in laminar flow is greater the greater the viscosity as is shown by the equation:

$$\tau = v \, du/dz \tag{14}$$

in which v is the kinematic viscosity, u is the velocity and z the height above the bed. The shear stress (τ) has the units newtons/square metre (Nm^{-2}). The flow reaches equilibrium with the boundary to produce a parabolic velocity profile in laminar flow. In turbulent flow the transfer of momentum incorporates not only the viscous effects but also the transfer of mass and momentum as 'parcels' of water move between the layers. This is usually accommodated in considering boundary layer flow by the eddy viscosity coefficient as a measure of the eddies moving up through the water. This eddy viscosity is also assumed to be proportional to both the velocity gradient and the depth, and so the shear is given by the equation:

$$\tau = (k.z)^{2.0} \, (du/dz)^2 \tag{15}$$

Manipulation reveals that with turbulent flow the velocity profile is logarithmic with respect to depth. In either condition, the velocity profile can be used in steady flow to indicate the energy transfer which is taking place between different levels and particularly at the boundary where the bed shear stress is assumed to provide energy for erosive work. As we shall see below the velocity profile and its associated energy transfer are also considered to be critical in sediment transport.

For a smooth wall, the boundary shear stress is being applied to overcome the drag and represents the energy losses. Since the drag is proportional to the square of the velocity one can be used to determine the other provided that the coefficient of proportionality is known. This forms the basis of formulae and coefficients such as the Chezy and Darcy–Weisbach formulae for determining the velocity of uniform steady flows in pipes. For laminar flow, where the boundary is actually stationary, the coefficients must by definition be independent of the roughness of the pipe wall and dependent only on the Reynolds number and decreases as the Reynolds number increases. With turbulent flow, however, the coefficients are independent of the Reynolds number and heavily dependent on the boundary roughness. Then they are often used to indicate different degrees of boundary roughness. In these cases the velocity profile (and hence rate of energy transfer) depend also on the roughness. The Colebrook–White equation enables both cases to be taken together:

$$1/\sqrt{f} = -2 \log{(k/3.7d + 2.51/Re\sqrt{f})} \tag{16}$$

in which k is the effective boundary roughness in mm, Re the Reynolds number and d the flow depth. When k is zero, the first term in the brackets disappears and the smooth log profile case dominates. When the Reynolds number is very large, the second term is very small and most of the energy is expended in overcoming boundary roughness.

The extreme case of expenditure on skin resistance is that of gravel bed and bouldery mountain streams. For gravel bed rivers the roughness elements are not usually uniformly distributed across the section, the sediment on the bed varies in size, and there are significant variations in cross-sectional form which have to be accommodated. Hey (1979) provides a modified version of the Colebrook–White equation which yields close estimates of the flow resistance when compared with measured field values provided that the assumptions are properly met, particularly that the flow is uniform.

When the flow is unsteady and non-uniform, the skin resistance energy losses are augmented by other forms of energy expenditure. These arise from changes in channel cross-section and depth, obstructions such as high flow deposits, changes in discharge at tributary junctions and phenomena such as waterfalls. Similarly on hillslopes the range of such phenomena is large and hitherto relatively little studied. In so far as many of them use energy which might otherwise be expended in sediment transfer or erosion, they are very important. In the design of soil conservation structures the dissipation of energy in non-uniform unsteady flows is of paramount importance. One source of energy expenditure is in overcoming internal distortion resistance. This generally arises when the flow is deflected away from the boundary by sharp changes (separation) which give rise to eddies, secondary circulation and increased shear rate. Leopold *et al.* (1960) conducted flume experiments using simple meandering channels and showed that the internal distortion resistance could amount to as much as three to four times the skin resistance and that it is dependent on the width to curvature ratio. When width is large relative to curvature, the bends become very 'tight' and the internal distortion component very large. They also described energy dissipation due to 'spill' resistance. This arises when water is elevated around a bend, and is forced to adjust to a steep downstream gradient in that direction rather by gradual upstream compensation in flow characteristics as a result of relatively high Reynolds numbers. With internal distortion resistance energy expenditure is again proportional to the square of velocity, though the coefficients are larger, whereas with spill resistance the effect sets in at a critical Reynolds number and is not a linear function of the square of velocity. The increase in energy transformation in a curved channel is also due, of course, to internal fluid friction arising from transverse circulation. Because of this extra resistance, meandering channels tend to be steeper than the corresponding straight channel. Chang (1983) argues that in addition to the width–curvature ratio, energy losses due to curvature are also directly proportional to the Froude number, the depth–radius ratio and the channel roughness, and confirms that the energy losses due to channel curvature may be as great as those due to skin friction. Even stronger effects were found by Ackers (1981).

Other important sources of energy dissipation arise from channel constrictions, both lateral and vertical, from vegetation within the channel and on the banks and from backwater effects among many others, even when there is no sediment movement on the boundary. It is therefore clear that while the major sources of energy dissipation can be identified (Bathurst, 1982) researchers in hydraulics are still far from developing an adequate resistance equation for the conditions prevailing in most natural streams.

3.4.3 Stream power

The Bernoulli approach, the resistance equation approach and boundary layer theory have provided the main staple of the practical approach to channel energetics. By contrast, in recent years attention has turned to the rate of energy expenditure especially in the context of sediment transport, although the general principles were outlined in the work of Bakhmeteff and Allen (1945). The rate of energy expenditure is called power and has the units of watts (joules s^{-1}) or, as unit power, watts per square metre (W/m^2). For two points along a profile the difference in potential energy is $\varrho g Q(h1-h2)$ where $h1$ and $h2$ are the respective heights, Q the discharge and ϱ the density of the flow. Dividing through by the horizontal difference yields power per unit length of the form:

$$\omega = \varrho g Q s \tag{17}$$

with s the stream bed slope and the units in $W\,m^{-1}$. The power per unit area, given by $\tau.u$, where τ is the bed shear stress and u the mean velocity, is the more commonly used of the two alternative expressions because it has the benefit of virtually eliminating scale effects. For British rivers, Ferguson (1981) has mapped both versions and the patterns are broadly similar (Figure 3.10) with unit stream power values in excess of $100-1000\,W\,m^{-2}$ for mountain streams in northern Britain, and values in the range $1-10\,W\,m^{-2}$ for streams in eastern Britain. A third version of stream power is that given by Yang (1972) for one-dimensional flow in which the average rate of energy expenditure is written as the average head loss per unit of time, that is:

$$dY/dt = (dx/dt)\,(dy/dx) \tag{18}$$

Since dy/dx is slope and dx/dt velocity, then the unit stream power per unit weight of water is given by $u.s$ which, he argues, will be minimized in natural channels. This proposition was subsequently used (Yang and Song, 1979) to

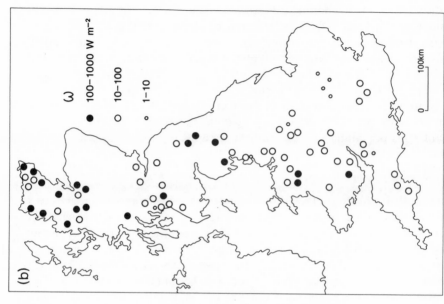

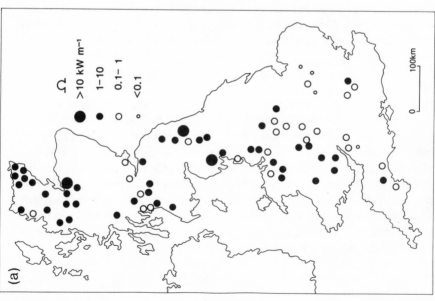

Figure 3.10   Stream power per unit length of channel (a) and per unit area of bed (b) for British streams at Bankfull (From Ferguson, 1981)

derive the parabolic velocity profile for laminar flow. For the turbulent case they deduce a linear profile in the laminar sublayer, a logarithmic law in the boundary layer, and parabolic profile in the outer flow zone. In their analysis the minimization hypothesis does not apply to unsteady behaviour, but the rate of energy expenditure should steadily decrease until steady flow prevails through adjustments in the channel morphology.

3.4.4 Hillslope runoff energetics

These different approaches to energy expenditure have been developed for pipes in the first instance and successively applied to man-made structures and later to open channel flow in natural rivers under conditions far from those for which they were developed. We consider finally their application to hillslope flow conditions. This is particularly difficult because the flow is often very irregular, it occurs usually when rain is falling, it is difficult if not impossible to estimate depths and velocities in the field, and vegetation plays a special part in the energy transformation considerations. Once again the dominant approach has been through friction factors as indicators of the resistance to flow and hence the relative energy expenditure in overcoming that resistance, though in recent years there have been some measurements of the actual energy applied by overland flow. The resistance coefficients most commonly used has been the Darcy–Weisbach resistance factor (K) which for laminar flow in pipes has a linear inverse relationship with the Reynolds number of the form 96/Re. For laminar flow over rough surfaces it takes the general form K/Re with $K > 96$. This general relationship appears to hold only up to about Re = 648 on a 1 per cent slope and Re = 140 on a 10 per cent slope with K increasing more and more as the slope becomes steeper and the ground rougher. With rainfall the K value is greater and thought to be a function of rainfall intensity. Typically Reynolds numbers on hillslopes lie in the range 200–16 000, with laminar flow predominating. The value of K may be determined in the laboratory as in rivers by estimating from velocity and depth data. In the field it has to be derived indirectly from the steady state storage, the rising hydrograph or by optimization of computed discharge curves against the observed (Woolhiser, 1975; Scoging and Thornes, 1980). Woolhiser tabulates values for unvegetated surfaces in the range 24–500, and for vegetated surfaces in the range 1000–40 000 indicating the enormous energy absorbing capacity of the vegetal cover. With rainfall, Emmett (1978) found that while the Reynolds numbers are typically in the laminar range, the flow has many of the characteristics of turbulent flow and used the term disturbed flow to describe these conditions.

The momentum of the falling rain is transferred to the overland flow and has a dominant vertical component compared with the downslope component of the latter. The net effect is to increase the depth of flow and hence absorb some of the energy. Emmett found that the friction factor in laboratory experiments was doubled in the presence of simulated rainfall. In the field this effect is greatly enhanced by vegetation and topography and there was a 10-fold increase in the energy required to overcome the vegetation cover. As in rivers the resistance to flow and hence energy expenditure are adjusted to energy availability in order to maintain an equilibrium and continuity in the slope form–hydraulic relationships.

3.5 RUNOFF EROSION

3.5.1 The classical approach

Most problems in erosion can be divided into three important categories, the erosion of solid rock by cavitation, the erosion of unconsolidated non-cohesive sediments, and the erosion of cohesive sediments. The second of these has been well studied, while the other two are still imperfectly understood in natural situations. However, they all have in common the application of forces in overcoming resistance. In the first and second cases the force is mechanical, in the third chemical forces are also being applied (compare Chapter 5, p. 119). In hillslope and channel erosion it is usually assumed that the work performed is in the application of normal or shear stresses to particles making up the boundary of the flow. These have previously been made available to the flow by weathering (in the case of hillslopes) or by mass movement, previous desposition or hillslope supply in the case of channels. One can recall that stress is force per unit area ($N\,m^{-2}$), and the work done is the product of the force and the distance over which it is applied ($N\,m^{-1}$); power is the rate at which the work is performed.

In a river erosion is measured by the net output of sediment from a reach (see Chapter 4, p. 97). This means that the balance of input and output must be negative and that more sediment must be entrained and transported than is deposited in the reach. For deposition (or settling) the force is provided by the immersed particle weight and resisted by form drag and skin friction (Shapiro, 1961). If we have a group of particles suspended in a fluid by an upward jet, and the particles are stationary, then the kinetic energy required to keep them there is the same as the potential energy consumed by falling particles in overcoming the resistances. For the falling particles, the relationship is:

$$V^2 = k(1-c)^2(\varrho-\gamma)gd/Cx$$

with c being the concentration of the particles (expressed as a fractional porosity), k a constant, ϱ and γ the density of the water and the particles respectively, Cx a shape factor, and d the diameter of the particles. This has the units $m^2\,s^{-2}$, and by multiplying the mass of the particles (kg) we obtain their kinetic energy. For entrainment the forces tending to move the particle must just exceed those tending to hold it to the bed of the flow. These forces are usually equated to provide a threshold-of-movement criterion, such as the Shields entrainment function whose derivation is given in Embleton and Thornes (1979). Classical work on suspended load transport has little to say about energetics, because the main model uses a diffusion assumption in which the upward movement of particles is assumed to be proportional to their concentration gradient. By contrast, physically defined (as opposed to empirical) bedload transport formulae again resort to stress arguments, the most popular being the tractive force type formulae.

3.5.2   Energy specific formulations

In 1966 R. A. Bagnold published a general theory of sediment transport based specifically on consideration of the stream energetics based on many years of study of sediment transport phenomena. He pointed out that the work done per unit of time in transporting sediment (that is, the power or rate of energy expenditure) is the immersed weight of the sediments being transported per unit area of bed multiplied by the transport velocity. The former is the immersed weight multiplied by the dynamic friction coefficient, that is:

$$(\varrho - \gamma)/mg \ \tan\theta u_b \tag{18}$$

where m is the bedload mass per unit area, $\tan \theta$ the dynamic friction coefficient, and u_b the bedload velocity. Now, as pointed out earlier, the available power per unit area of bed due to the flow is $\tau_0 v$ where τ_0 is the bed shear stress and v the fluid velocity. Not all the energy available can be used since, as we showed earlier, much of it is expended in overcoming fluid resistance and the efficiency e_b is the ratio of actual work rate to available power, so:

$$(\varrho - \gamma)/\gamma mg \ \tan\theta \ ub = \tau_0 v e_b$$

$\gamma \tau_0 . v$ is also equal to γqs, where q is the discharge per unit width and s the slope. This formulation is particularly useful for slope work. The rate of sediment transport is therefore theoretically predictable knowing the efficiency. Below the threshold for movement this is zero; above it stream power or some version of it is expected to correlate well with sediment transport rate. The importance of this formulation is in pointing to the appropriate conceptualization of the process and thereby identifying the parameters which ought best to define the transport rate. Since v^3 is equivalent to power, this is sometimes used as a substitute and correlated with sediment transport rates, but Bagnold (1977) also incorporates a threshold w_{crit} which implies (Allen, 1985) a predictive equation of the general form:

$$(v_m^2 - v_{crit}^2)\,(u_m - u_{crit})$$

Leopold and Emmett (1977) have shown that e_b *is constant for* $w > w_{crit}$ and is lower for coarser bed material, and Bagnold (1977) shows that efficiency for bedload transport is strongly conditioned by the depth/width ratio of the flow (Figure 3.11). As the stream gets shallower for a fixed width the efficiency of energy utilization increases. This may explain why in ephemeral channels there is a progressive widening of the channel to accommodate increased bedloads until some critical value is reached after which there is a rapid contraction in width again (Thornes, 1977). Bedload transport efficiencies may be as high as 10–15 per cent which largely explains the relative efficiency of gullies and streams in erosion when compared with splash. The other reason, of course, is because the actual unit energy applied is much higher.

Bagnold (1966) provided a comparable analysis for the transport of suspended and wash load so that, unlike the classical analysis, in which the bedload, suspended load and wash load are treated separately, the Bagnold transport theory is completely general.

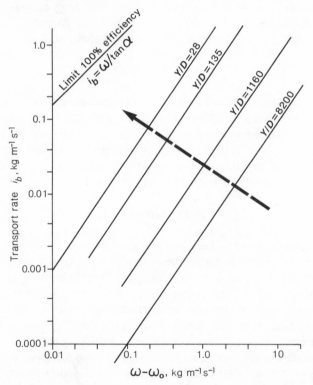

Figure 3.11   The relationship between bedload efficiency and width to depth ratios (After Bagnold, 1977)

There have been many works in recent years which have tested the use of the Bagnold stream power principle in accounting for observed sediment transport in channels and on hillslopes. In the latter, the main difficulty, as we pointed out in the introduction, is that erosion rates are not only a function of energy or power available for transport (that is, the capacity) but also of the availability of material. Morgan (1980a) found that none of the five sediment transport equations he tried, including one based on kinetic energy of the flow and several involving stream power terms, performed well in accounting for the sediment transport observed. More recently Govers (1985) found that in thin flows related to rilling, sediment concentration ($mg\,l^{-1}$) correlated very well with the Bagnold stream power expression, as did the median grain size. This latter observation suggests that flow competence as well as capacity may be related to the rate of energy expenditure at least in non-cohesive (that is, not detachment-limiting) materials. Schick and Lekach (1981) found that the 'empirical' version of Bagnold's (1977), $s = g(w-wc)^{2/3}$, provided a reasonable model for very high rates of bedload transport in Wadi Mikeimin, and Leopold and Emmett (1977) confirmed its applicability in one of the most elaborate observations of bedload transport ever performed. The energy formulation for sediment transport therefore seems to be gaining ground with natural scientists though it is still not widely used by engineers.

3.5.3 Unified transport theory

We have seen that the transport of material has been regarded as a set of disjoint processes, and the Bagnold theory attempts to link together the theory of wash, suspended and bedload fluvial transport. In recent years Culling (1983) has set out to provide a unified basis for all transport phenomena ranging from soil creep to fluvial transport based on rate process theory. A full treatment is contained in Culling (1985) and only the briefest of sketches can be provided here.

The basic idea is that molecules are subject to activating energies that cause them to oscillate, enabling them to pass over constraints or barriers (for example in a crystal structure) and thus changing the arrangement of the particulate medium. The success (and hence transport) rate depends on the spacing and heights of the barriers and the magnitude of the activating energies as shown diagrammatically in Figure 3.12. By knowing the probability distributions of the movements due to the activation and the distribution of the 'heights' it is possible theoretically to derive the distribution of successes and hence the rearrangement of the medium. One can envisage the process as being like a set of peas on a corrugated board which is being constantly tapped underneath. The activation energy is provided by the random tapping and the ridges are at random heights. The transport rate could then be observed by comparing the number of peas redistributed from a line in the middle to slots further and

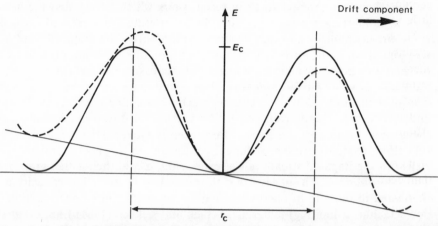

Figure 3.12   The barrier concept of rate process theory. The average spacing between barriers is γ^2

further from the centre. Superimposing a tilt on the board has the effect of lowering the barriers to movement on the downslope side and hence increasing the drift in this direction. In soil creep for which the theory is fully developed, transformation has been made to consider particulate as well as molecular elements and activation is provided by physical, chemical, floral and faunal agencies which displace the particles at random. Gravity is a ubiquitous, persistent and reasonably constant but weak force, so Culling emphasizes the Brownian or stochastic component relative to the drift or gravitational component and the process is therefore diffusive. The distribution of 'barriers' is related in the case of soil creep to the distribution of void spaces and to the various forms of bonding between the particles. So far the theory is still in its infancy and has undergone only limited testing (Flavell, 1985), but in principle the proposed unified theory is very attractive and re-emphasizes the importance of general energetics in geomorphic processes.

3.6   ENERGETICS APPLIED

In this final section we wish to stress the point that although our discussion has been essentially academic, the principles discussed form the basis for most soil conservation practices. There is a huge body of hydraulics literature covering the design of river channel structures for reducing energy (see, for example, Elevatorski, 1959). For hillslope erosion under agricultural conditions the discussion provided by Morgan (1980b) and the references therein provide an excellent starting point. The implementation of conservation schemes is very much a matter of socioeconomic and cultural as well as mechanical and geomorphological considerations, but without the latter the former can be of little value.

The basic mechanisms of soil conservation involve either a reduction of the available energy or power of the rainfall or flow, or an increase in the rate of energy expenditure through surface control. In practice both are often achieved through one agency so the division is rather arbitrary. Rainfall energy reduction is generally achieved through an increase of the cover. As we and others have shown, the practice of retaining forest cover while removing the scrub and ground layer may be rather misguided since it is the ground and near ground cover that provided the capacity for energy absorption without enhancement of its erosive potential. It is likewise true that row crops may increase rather than decrease the energy available for erosion. By contrast, native grass covers, secondary shrub growth and artificial mulching usually not only dissipate the rainfall energy but also provide higher interception and infiltration levels which reduces runoff. Many artificial practices specifically aim at increasing infiltration losses by increasing the surface permeability by ploughing and chemical treatment. In conventional irrigation practice a compromise has to be reached between providing sufficient energy to keep the water moving to ensure a uniform and adequate supply to the further parts of the field on the one hand, while reducing scour and providing sufficient infiltration at each point on the other. This delicate balance is achieved by adjusting the field slope, the field roughness and/or the rate of discharge supply. In primitive agriculture achieving this energy balance may itself involve the expenditure of considerable effort, and in modern farming drip irrigation is only now beginning to provide an alternative solution to this complex and expensive problem. Despite the considerable progress made in conservation energetics, combining the relevant equations is a complex process (see, for example, the work of Smith (1972) on border flood irrigation) and obtaining realistic solutions for design purposes is only possible in very limited conditions. This point is well illustrated in the design of regime canals for irrigation systems (Withers and Vipond, 1974; Thornes, 1979).

The management of gully systems is also a problem in applied erosional energetics. In the event that watershed management has failed or been implemented too late, conservation efforts often have to start in the gully or incipient channel. S. A. Schumm and his coworkers (see, for example, Schumm and Beathard, 1976) have shown particularly that it is possible to predict the slope conditions in the power equation which lead to gully initiation by regional geomorphological analysis, while Kirkby (1980) has approached the problem through the discharge term. Once initiated, propagation of the gully head is a function of the concentration of power through discharge and slope in the gully head area in addition to the resistance of the materials to the applied stress (Thornes, 1984). Most effort, however, has been concerned with the transformation of energy through the design and construction of check dams,

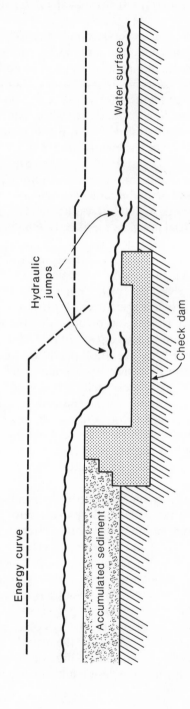

Figure 3.13   Cross section through a structure designed to check sediment movement and to provide for energy dissipation through hydraulic jumps

which reduce the channel slope by aggradation, the discharge by infiltration, and dissipate energy by providing a hydraulic jump in the design of the structure. An example of this is shown in Figure 3.13. Geomorphic assessment, and analysis of rainfall and runoff energy and power coupled with engineering design for energy dissipation, can provide a powerful combination for the solution of the problems of erosional energetics within catchments.

References

Ackers, P. (1981). Meandering channels and the influence of bed material. In Hey, R. D., Bathurst, J. C., and Thorne, C. R. (eds) *Gravel Bed Rivers*, Wiley: Chichester, 389–422.

Ahnert, F. (1970). Functional relationships between denudation relief and uplift in large mid-latitude drainage basins, *American Journal of Science*, **268**, 243–263.

Al-Durrah, M. and Bradford, J. M. (1982). New methods of studying soil detachment due of raindrop impact, *Journal of the Soil Science Society of America*, **45**, 949–953.

Allen, J. R. L. (1985). *Principles of Physical Sedimentology*. George Allen and Unwin: London, 272.

Bakhmeteff, B. A. and Allen, W. (1945). The mechanism of energy loss in friction, *American Society of Civil Engineers*, **71**, 129–166.

Bagnold, R. A. (1966). An approach to the sediment transport problem from general physics, *Professional Paper*, United States Geological Survey, 422-I, 37.

Bagnold, R. A. (1977). Bed load transport by natural rivers, *Water Resources Research*, **13(2)**, 303–312.

Bathurst, J. C. (1982). Theoretical aspects of flow resistance. In Hey, R. D., Bathurst, J. C., and Thorne, C. R. (eds) *Gravel Bed Rivers*, Wiley: Chichester, 83–108.

Best, A. C. (1950a). The size distribution of raindrops, *Quarterly Journal of the Royal Meteorological Society*, **76**, 16–36.

Best, A. C. (1950b). Empirical formulae for the terminal velocity of water drops falling through the atmosphere, *Quarterly Journal of the Royal Meteorological Society*, **76**, 302–311.

Bollinne, A. (1978). Study of the importance of splash and wash on cultivated loamy soils of Hesbaye (Belgium), *Earth Surface Processes*, **3**, 71–84.

Brandt, C. J. (1986). *Transformation of the kinetic energy of rainfall with variable tree canopies*, unpublished PhD thesis, University of London.

Bubenzer, G. D. and Jones, B. A. (1971). Drop size and impact velocity effect on the detachment of soils under simulated rainfall, *Transactions of the American Society of Agricultural Engineers*, **14(4)**, 625–628.

Caine, N. (1976). A uniform measure of sub-aerial erosion, *Bulletin of the Geological Society of America*, **87**, 137–140.

Chang, H. H. (1983). Energy expenditure in curved open channels, *Journal of Hydraulic Engineering*, **109**, 1012–1023.

Chapman, G. (1948). Size of raindrops and their striking force at the soil surface in a red pine plantation, *Transactions of the American Geophysical Union*, **29**, 664–670.

Chorley, R. J., Schumm, S. A., and Sugden, D. (1985). *Geomorphology*. Methuen: London, 607.

Culling, W. E. H. (1983). Rate process theory of geomorphic soil creep, *Catena*, supplement 4, 191–214.

Culling, W. E. H. (1985). Towards a unified theory of particulate flow in geomorphic processes, *Royal Holloway and Bedford New College, Department of Geography, Discussion Paper*, 1, 125.

Davies, C. N. (1942). *Unpublished Ministry of Supply reports*, quoted by Sutton in Air Ministry report, MRP No. 40.

Dohrenwend, R. E. (1977). *Raindrop Erosion in the Forest*. Michigan Technological University, Ford Forestry Center: L'Anse Michigan 49946, Research Note 24, 19.

Ekern, P. G. (1950). Raindrop impact as the force initiating soil erosion, *Proceedings of the Soil Science Society of America*, 15, 7–10.

Ekern, P. G. (1953). Problems of raindrop impact erosion, *Agricultural Engineering*, 34(1), 23–25.

Elevatorski, E. A. (1959). *Hydraulic Energy Dissipators*. McGraw-Hill: New York, 236.

Ellison, W. E. (1944). Studies of raindrop erosion, *Agricultural Engineering*, 25(4), 131–136, (5), 181–182.

Embleton, C. E. and Thornes, J. B. (1979). *Process in Geomorphology*. Edward Arnold: London, 436.

Emmett, W. W. (1978). Overland flow. In Kirkby, M. J. K. (ed.) *Hillslope Hydrology*. Wiley: Chichester, 145–170.

Ferguson, R. I. (1981). Channels forms and channel changes. In Lewin, J. (ed.) *British Rivers*, George Allen and Unwin: London, 90–121.

Flavell, W. S. (1985). *Field verification of a stochastic model of soil creep*, unpublished PhD thesis, University of London, 197.

Flower, W. D. (1928). The terminal velocity of drops, *Proceedings of the Physical Society of London*, 40, 167–176.

Foster, G. R. and Meyer, L. D. (1972). A closed-form soil erosion equation for upland areas. In Shen, H. W. (ed.) *Sedimentation*, H. W. Shen: Fort Collins, Colorado, 12.1–12.19.

Ghadiri, H. and Payne, D. (1979). Raindrop impact and soil splash. In R. Lal and D. J. Greenland (eds) *Soil Physical Properties and Crop Production in the Tropics*. Wiley-Interscience: Chichester, 95–104.

Govers, G. (1985). Selectivity and transport capacity of thin flows in relation to rill erosion, *Catena*, 12, 35–49.

Gunn, R. and Kinzer, G. D. (1949). The terminal velocity of fall for water droplets in stagnant air, *Journal of Meteorology*, 6, 243–248.

Hall, M. J. (1970). Use of the stain method in determining the drop-size distribution of coarse liquid sprays, *Transactions of the American Society of Agricultural Engineers*, 13, 33–37, 41.

Harlow, F. H. and Shannon, J. P. (1967). The splash of a liquid drop, *Journal of Applied Physics*, 38(10), 3855–3866.

Hey, R. D. (1979). Flow resistance in gravel-bed rivers, Proceedings of the American Society of Civil Engineers, *Journal of the Hydraulics Division*, 105, 365–377.

Horton, R. E. (1948). Statistical distribution of drop sizes and the occurrence of dominant drop sizes in rain, *Transactions of the American Geophysical Union*, 29(5), 624–630.

Houze, R. A., Hobbs, P. V., Parsons, D. B., and Hertzegh, P. H. (1979). Size distribution of precipitation particles in frontal clouds, *Journal of the Atmospheric Sciences*, 36(1), 156–162.

Huang, C., Bradford, J. M., and Cushman, J. H. (1982). A numerical study of raindrop impact phenomena: the rigid case, *Journal of the Soil Science Society of America*, 46, 14–19.

Hudson, N. W. (1971). *Soil Conservation*. Batsford: London, 320.

Jansson, M. B. (1982). Land erosion by water in different climates, *UNGI Rapport No. 57*, Department of Physical Geography: Uppsala University, 151.

Kanen, J. D. (1986). *Applied Hydraulics*. Holt, Reinhart and Winston: New York, 513.

Kinnel, P. I. A. (1973). The problem of assessing the erosive power of rainfall from meteorological observations, *Proceedings, Soil Science Society of America*, **37(4)**, 617–621.

Kinnel, P. I. A. (1981). Rainfall intensity–kinetic energy relationships for soil loss prediction, *Journal of the Soil Science Society of America*, **45**, 153–155.

Kirkby, M. J. (1980). The stream head as a significant geomorphic threshold. In Coates, D. R. and Vitek, J. (eds) *Thresholds in Geomorphology*, George Allen and Unwin: London, 53–74.

Klett, J. D. (1971). On the break up of water in air, *Journal of the Atmospheric Sciences*, **28**, 646–647.

Kneale, W. R. (1982). Field measurements of rainfall drop-size distribution and the relationship between rainfall parameters and soil movement by rain splash, *Earth Surface Processes and Landforms*, **7**, 499–502.

Laws, J. O. (1941). Measurements of the fall-velocities of waterdrops and raindrops, *Transactions of the American Geophysical Union*, **22**, 709–712.

Laws, J. O. and Parsons, D. A. (1943). The relation of raindrop size to intensity, *Transactions of the American Geophysical Union*, **24**, 452–459.

Lenard, P. (1904). Uber Regen, *Meteorologische Zeitschrift*, **21**, 248–262.

Leopold, L. and Emmett, W. W. (1977) (1976). Bedload measurements, East Fork River, Wyoming, *Proceedings Natural Academy of Sciences USA*, **74**, 2644–2648.

Leopold, L., Bagnold, R. A., Wolman, M. G., and Brush, L. (1960). Flow resistance in sinuous or irregular channels, *Professional Paper*, United States Geological Survey, Washington, 282-D, 111–134.

Marshall, J. S. and Palmer, W. McK. (1948). Relation of raindrop size to intensity, *Journal of Meteorology*, **5**, 165–166.

Mason, B. J. (1957). *The Physics of Clouds*. Clarendon Press: Oxford, 481.

Mason, B. J. and Andrews, J. B. (1960). Drop-size distribution from various types of rain, *Quarterly Journal of the Royal Meteorological Society*, **86**, 346–353.

Mason, B. J. and Ramandham, R. (1953). A photoelectric spectrometer, *Quarterly Journal of the Royal Meteorological Society*, **79**, 490–495.

McGregor, K. C. and Mutchler, C. K. (1978). The effect of crop canopy on raindrop size distribution and energy, *USDA Sedimentation Laboratory Annual Report*. Oxford, MS, 38655 USA.

Meyer, L. D. and Wischmeier, W. H. (1969). Mathematical simulation of the process of soil erosion by running water, *Transactions of the American Society of Agricultural Engineers*, **12**, 754–758.

Mihara, Y. (1951). Raindrops in soil erosion (English translation) *National Institute of Agricultural Science*, Bulletin No. 1 Tokyo, Japan, 59.

Mitchell, J. K. and Bubenzer, G. D. (1980). Soil loss estimation. In Kirkby, M. J. and Morgan, R. P. C. (eds) *Soil Erosion*, Wiley: Chichester, 17–62.

Morgan, R. P. C. (1978). Field studies of rainsplash erosion, *Earth Surface Processes*, **3(3)**, 295–299.

Morgan, R. P. C. (1980a). Field studies of sediment transport by overland flow, *Earth Surface Processes and Landforms*, **5(4)**, 307–316.

Morgan, R. P. C. (1980b). Implications. In Kirkby, M. J. and Morgan, R. P. C. (eds) *Soil Erosion*, Wiley: Chichester, 253–301.

Morgan, R. P. C. (1982). Splash detachment under plant covers: results and implications of a field study, *Transactions of the American Society of Agricultural Engineers*, **25(4)**, 987–991.

Morris, S. E. (1986). The significance of rainsplash in the surficial debris cascade of the Colorado Front Range foothills, *Earth Surface Processes and Landforms*, **11**, 11–22.

Mosley, M. P. (1982). The effect of a New Zealand beech forest canopy on the kinetic energy of water drops and on surface erosion, *Earth Surface Processes and Landforms*, **7**, 103–107.

Mutchler, C. K. (1967). Parameters for describing raindrop splash, *Journal of Soil Water Conservation*, **22(3)**, 91–94.

Noble, C. A. and Morgan, R. P. C. (1983). Rainfall interception and splash detachment with a brussels sprout plant: a laboratory simulation, *Earth Surface Processes and Landforms*, **8**, 569–577.

Ovington, J. D. (1954). A comparison of rainfall in different woodlands, *Forestry*, **27**, 41–53.

Palmer, R. (1963). Waterdrop impactometer, *Agricultural Engineering*, **44**, 198–199.

Palmer, R. S. (1965). Waterdrop impact forces, *Transactions of the American Society of Agricultural Engineers*, **8(1)**, 70–72.

Park, S. W., Mitchell, J. K., and Bubenzer, G. D. (1982). Splash erosion modelling: physical analysis, *Transactions of the American Society of Agricultural Engineers*, **25**, 357–361.

Pearce, A. J. (1976). Magnitude and frequency of erosion by Hortonian overland flow, *Journal of Geology*, **84**, 65–80.

Pruppacher, H. R. and Pitter, R. L. (1971). A semi-empirical determination of the shape of cloud and rain drops, *Journal of the Atmospheric Sciences*, **28**, 86–94.

Quinn, N. W. and Laflen, J. M. (1981). Properties of transformed rainfall under corn canopy, *American Society of Agricultural Engineers* paper No. 81–2059, 15.

Riezebos, H. Th. and Epema, G. F. (1985). Drop shape and erosivity. Part II: Splash detachment, transport and erosivity indices, *Earth Surface Processes and Landforms*, **10**, 69–74.

Schick, A. P. and Lekach, J. (1981). High bedload transport rates in relation to stream power, Wadi Mikeimin, *Catena*, **8**, 43–47.

Schindelhaur, (1925). Versuch einer Registrierung der Tropfenzahl bei Regenfällen. *Meteorologische Zeitschrift*, **42**, 25.

Schottman, W. R. (1978). *Estimation of the penetration of high energy raindrops through a plant canopy*, unpublished PhD thesis, Cornell University, Ithaca, NY.

Schumm, S. A. and Beathard, R. M. (1976). Geomorphic thresholds: an approach to river management. In *Rivers '76*, American Society of Civil Engineers, 707–724.

Scoging, H. M. and Thornes, J. B. (1980). Infiltration characteristics in a semi-arid environment, *International Association of Science Hydrology, Publication*, **128**, 159–168.

Shapiro, A. H. (1961). *Shape and Flow*. Heinemann: London, 185.

Smith, R. E., (1972). Border irrigation advance and ephemeral flood waves, *Journal of the Irrigation Division*, American Society of Civil Engineers, **98**, 128–145

Sreenivas, L., Johnston, J. R., and Hill, H. O. (1947). Some relationships of vegetation and soil detachment in the erosion process, *Proceedings, Soil Science Society of America*, **11**, 474–479.

Srivastava, R. C. (1971). Size distributions of raindrops generated by their break up and coalescence, *Journal of the Atmospheric Sciences*, **28**, 410–415.

Thornes, J. B. (1977). Channel changes in ephemeral streams: observations, problems and models. In Gregory, K. J. (ed.) *River Channel Changes*, Wiley, 317–335.

Thornes, J. B. (1979). Fluvial process. In Embleton, C. E. and Thornes, J. B. (eds) *Process in Geomorphology*, Edward Arnold: London, 213–271.

Thornes, J. B. (1984). Gully growth bifurcation. In *Erosion Control: Man and Nature*. International Erosion Control Association: Denver, 131–140.

Tsukamoto, Y. (1966). Raindrops under forest canopies and splash erosion, *Bulletin of Experimental Forestry, Tokyo University of Agricultural Technology*, **5**, 65–77.

Wiersum, K. F. (1985). Effects of various vegetation layers of an *Acacia auriculiformis* forest plantation on surface erosion at Java, Indonesia. In El-Swaify, S. A., Moldenhauer, W. C., and Lo, A. (eds) *Soil Erosion and Conservation*, Soil Conservation Society of America, 79–89.

Wischmeier, W. H. and Smith, D. D. (1958). Rainfall energy and its relationship to soil loss, *Transactions of the American Geophysical Union*, **29**, 285–291.

Withers, B. and Vipond, S. (1974). *Irrigation: Design and Practice*, Batsford: London, 306.

Woolhiser, D. (1975). Simulation of unsteady overland flow. In Mamood, K. and Yevjevich, V. (eds) *Unsteady Flow in Open Channels*, Water Resources Publications: Fort Collins, Colorado, 485–508.

Yang, C. T. (1972). Unit stream power and sediment transport, *Journal of the Hydraulics Division*, American Society of Civil Engineers, **98**, 1805–1826.

Yang, C. T. and Song, C. C. S. (1979). Velocity profiles and minimum stream power, *Journal of the Hydraulics Division*, American Society of Civil Engineers, 105(HYH), 981–998.

Energetics of Physical Environment
Edited by K. J. Gregory
© 1987 John Wiley & Sons Ltd

4

Rainfall, Runoff and Erosion of the Land: A Global View

D. E. WALLING

Department of Geography, University of Exeter

4.1 THE GLOBAL HYDROLOGICAL CYCLE

The global hydrological cycle is closely linked to the global energy budget and provides a classic example of a closed system. Solar radiation provides the energy for evaporation from the oceans and for evaporation and transpiration (evapotranspiration) from the land surface of the earth, and the water vapour produced is dispersed upwards into the atmosphere where it circulates. This in turn provides the source of moisture for precipitation which replenishes the water lost by evapotranspiration. Some of the water falling on the land is evaporated or transpired by vegetation, but the excess flows to the oceans, thereby completing the cycle. The significance of this cycle in the global energy budget and heat balance is emphasized when it is recognized that approximately 85 per cent of the radiation balance at the earth's surface is accounted for by evapotranspiration processes, with the remaining 15 per cent contributing to turbulent heat exchange with the atmosphere (UNESCO, 1978).

It is difficult to provide an accurate assessment of the precise volumes of water involved in the global hydrological cycle, since there are many areas of the continents for which measurements of precipitation and more particularly runoff and evapotranspiration are lacking and there are few direct measurements of precipitation and evaporation available for the oceans. However, a detailed analysis of the global water balance undertaken by Soviet scientists and published in 1974 (USSR National Committee for the International Hydrological Decade, 1974) provides what are probably the best currently available estimates. These indicate that on an annual basis, the total volume of precipitation received at the earth's surface is equivalent to $577 \times 10^3 \text{ km}^3$ or an average depth of 1130 mm. This is balanced by an equivalent volume of moisture released by evapotranspiration from the oceans and continents.

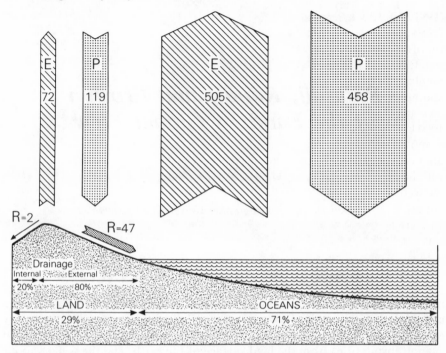

Figure 4.1 A schematic representation of the global hydrological cycle. Based on data in USSR National Committee for the International Hydrological Decade (1974). P: precipitation; E: evaporation or evapotranspiration; R: runoff; 505: value of mean annual flux (10^3 km^3)

Assuming an energy requirement of 2257 J g^{-1} for the conversion of water from liquid to vapour, the total energy input required to drive the global hydrological cycle is equivalent to 1.3×10^{24} J year^{-1}. The enormity of this value can perhaps be better appreciated if it is viewed as equivalent to the output of approximately 40 million major (1000 MW) power stations.

Figure 4.1 provides detail on the relative roles of the oceans and continents in the global hydrological cycle. The continents, which represent 29 per cent of the surface of the earth, account for 20 per cent of the annual precipitation input and 12.5 per cent of the annual evapotranspiration loss. The discrepancy between precipitation and evapotranspiration points to a water surplus which results in an annual runoff of 47×10^3 km^3 to the oceans. Some runoff to inland seas occurs, but this ultimately contributes to the evapotranspiration loss from the land surface.

Although the global hydrological cycle depicted in Figure 4.1 is necessarily dominated by natural forces and controls, it is important to recognize that man's increasing exploitation of the world's water resources must inevitably exert some influence on the processes involved. For example, it has been estimated

(UNESCO, 1978) that the total volume of water currently stored in man-made reservoirs amounts to about 5000 km^3 or about 11 per cent of the annual runoff from the land surface of the globe to the oceans. This represents a significant modification to the natural cycling of water between land and oceans. Similarly, the use of river water and groundwater for irrigation represents another disruption of the natural system. The total irrigated area in the world now amounts to some 300 million ha and involves a water use of approximately 4000 km^3. About 50 per cent of this water use is accounted for by increased evapotranspiration from the irrigated areas, which in turn represents a small, but nevertheless significant, modification to the natural cycle. Feedback effects will, however, operate since although increased water abstraction will result in reduced river runoff, the increased evapotranspiration and associated atmospheric moisture could result in increased local precipitation and therefore an increase in gross runoff. Although the volumes cited above represent only small proportions of the total quantities involved in the natural global cycle, it must be accepted that demands for water and associated water resource exploitation will inevitably intensify in the future and human impact on the global hydrological cycle must become increasingly significant. Proposals for major regional water transfers, such as the Soviet plans to divert water from northward flowing rivers towards the south, could clearly have an important effect.

4.2 THE EROSIONAL IMPACT

The global hydrological cycle exerts an important control on climate and water availability for human activity, but the continual cycling of water through the system also has important geomorphological implications. Precipitation striking the land surface and runoff across its surface possess considerable erosive energy and represent important agents of *mechanical* denudation. Furthermore, the vast volume of water associated with runoff from the continents to the oceans comes into contact with the soils and rocks of the land surface and removes considerable quantities of material in solution. This represents an agent of *chemical* denudation. Both result in erosion or denudation of the land surface of the globe and the transfer of material from the land to the oceans.

Looking in more detail at the processes of mechanical denudation, although necessarily at the level of global generalization, it is useful to distinguish between the erosive effects of precipitation striking the land surface and the erosion and transport of material associated with subsequent runoff from the land to the oceans. In the former case raindrop impact possesses considerable erosive energy, and Mihara (1959) has suggested that approximately two-thirds of this energy will be expended in forming an impact crater and moving soil particles, whilst the remainder is expended in the generation of spray. Particles may be splashed to heights of up to 30 cm and over distances of more than 90 cm. The kinetic energy

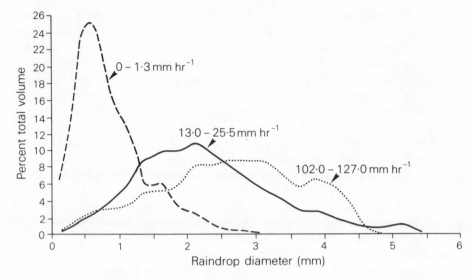

Figure 4.2 Rainfall energy characteristics. Typical drop-size distributions for storms of different intensities are illustrated as recorded in South-Central United States by Carter *et al.* (1974)

of raindrop impact is a function of the drop diameter, since the latter controls the fall velocity. Individual rainfall events will be characterized by a wide range of drop sizes, although the mean drop size commonly increases with increasing rainfall intensity (see, for example, Figure 4.2). The total kinetic energy associated with a unit depth of rainfall will therefore increase with increasing rainfall intensity (see, for example Figure 3.4, p.58). The drop size distribution associated with a particular rainfall intensity will, however, vary from place to place in response to variations in the conditions associated with precipitation generation, and the relationship between kinetic energy per unit depth of rainfall and rainfall intensity will accordingly show considerable variability (see Figure 3.4 and pp. 55–59).

The erosive effects of raindrop impact will be influenced by the degree of protection afforded to the soil surface by vegetation. Meyer and Wischmeier (1969) have also suggested that it is useful to distinguish two components of splash erosion, namely splash detachment and splash transport. Thus although large quantities of soil may be detached by splash from a flat field under bare conditions, no net loss will occur. On sloping areas, however, splash will generate a preferential downslope migration of soil particles (splash transport) and net loss or erosion may therefore occur. Meyer and Wischmeier (1969) suggest that splash transport will be directly proportional to slope angle.

The runoff component of the generalized representation of the global hydrological cycle depicted in Figure 4.1 refers essentially to water flowing in

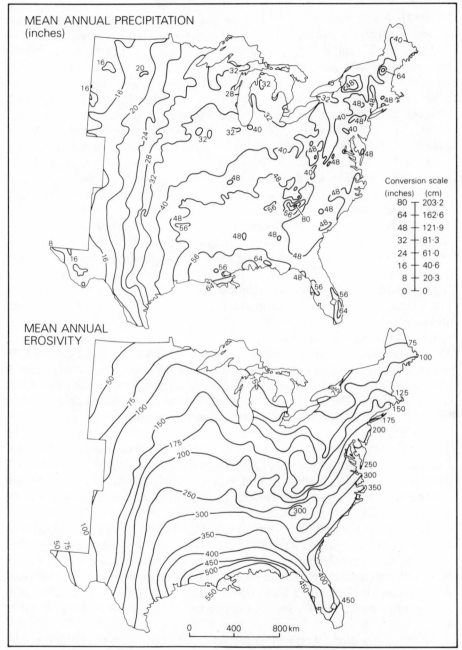

Figure 4.3 Comparison of mean annual precipitation and erosivity maps for central and eastern United States. The mean annual erosivity map utilizes the EI_{30} index, and is based on Wischmeier and Smith (1978)

rivers. In considering the erosive effects of runoff, one must, however, take account of water flowing across the entire land surface, including sheet flow and flow in rills and gullies as well as in river channels. Furthermore, it must be recognized that some of the water flowing in rivers will be supplied from subsurface groundwater sources, rather than surface origins, and that its erosive effects will therefore be limited primarily to the channel. Sheet flow generally behaves as laminar flow and typically evidences velocities in the range 2–30 cm s^{-1}. It therefore possesses limited erosive potential. When concentrated in rills and gullies, flow becomes turbulent, velocities increase (>30 cm s^{-1}) and both erosion and transport potential are greater. In larger river channels velocities may increase further, but the erosion potential is essentially restricted to the channel perimeter and channels frequently function primarily as transport routes.

In reality the distinction between the erosive effects of rainsplash and of surface runoff is of limited relevance, since the two processes act in combination. More particularly, rain splash will frequently play an important role in *detaching* soil particles which will subsequently be *transported* by surface runoff. The 'Universal Soil Loss Equation' (Wischmeier and Smith, 1978) which was developed in the United States as a tool for predicting the intensity of soil erosion by sheet and rill erosion on agricultural land under different cropping and management practices, provides a useful example of the interaction of the two processes. The erosive potential of rainfall and runoff is represented in this equation by the product of the total kinetic energy (E) and the maximum 30 min rainfall intensity (I_{30}) for a rainfall event. The former term reflects the energy of rain splash whilst the latter accounts for the incidence and magnitude of surface runoff, since this is influenced directly by rainfall intensity. Summation of the EI_{30} values for all storms occurring in a year provides an annual index of erosion potential or erosivity. Figure 4.3 depicts a generalized map of annual EI_{30} values for Central and Eastern USA and it is instructive to compare this with the equivalent map of mean annual precipitation. Whereas annual precipitation evidences a four-fold range across this region, erosion potential as represented by the EI_{30} index varies by more than an order of magnitude, emphasizing the importance of rainfall intensity as distinct from amount. Several workers in other areas of the world have also sought to define general indices representing the erosion potential or erosivity of rainfall and associated runoff by considering rainfall amount and intensity and its distribution during the year. Rainfall concentrated into a short period is more likely to generate substantial amounts of surface runoff. Several examples of these indices are listed in Table 4.1.

Although not directly related to the action of rain splash or running water, discussion of the role of water in mechanical denudation must also consider its significance to other processes, particularly mass movements. Water is an essential component of many rapid mass movement phenomena such as debris

Table 4.1   Indices of erosivity proposed by various workers

Index			Proposer
(i) p^2/P			
	Where: $p =$	mean monthly precipitation of wettest month	Fournier (1960)
	$P =$	mean annual precipitation	
(ii) $\sum_{1}^{12} pi^2/P$			
	Where: $pi =$	mean monthly precipitation	Arnoldus (1977)
	$P =$	mean annual precipitation	
(iii) $P.\sigma$			
	Where: $P =$	mean annual precipitation	Ciccaci, Fredi and Lupia Palmieri (1977)
	$\sigma =$	standard deviation of mean monthly precipitation totals	
(iv) $b.c$			
	Where: $b =$	number of days with precipitation > 30 mm	Demmak (1982)
	$c =$	percentage of annual precipitation falling on days with precipitation > 30 mm	
(v) $[\sum_{1}^{k} A.I_m]/100$			
	Where: $A =$	rainfall for an individual storm	Lal (1976)
	$I_m =$	maximum rainfall intensity of individual storm	
	$k =$	number of storms in year	

flows, debris slides, mudflows and landslides and these are frequently closely linked to fluvial processes. For example, material generated by mass movements may be an important source of the sediment transported by a river draining an area of steep unstable terrain. In general terms, increased precipitation and water availability must be seen as promoting mass movement activity as well as fluvial processes, particularly during high magnitude storm events.

Precipitation reaching the earth's surface cannot be viewed as pure water, since it contains small amounts of material in solution derived from atmospheric aerosols. However, the associated concentrations of dissolved material are commonly low (< 10 mg l^{-1}). Much higher concentrations are found in river water (for example, > 100 mg l^{-1}) and the associated increase can be largely related to the mobilization and removal of solutes as the water passes through the soils and rocks of the earth's surface. This removal of material in solution represents an important agent of *chemical* denudation. In some circumstances, such as when water passes through evaporite deposits, the mobilization of solutes may be viewed as a simple solution process, but in most cases chemical weathering processes are involved, with water playing an important role both in the weathering process itself and in removing the soluble products. These chemical weathering processes will frequently be aided by uptake of carbon dioxide and organic acids by water moving through the vegetation cover and into the soil and rock.

Space does not permit a detailed review of the many weathering reactions that occur, but two examples can be usefully cited. Since silicate minerals account for 70 per cent or more of the rocks in contact with the hydrological cycle (Bricker and Garrels, 1967), the first involves the dissolution or hydrolysis of aluminium silicates. In this case the primary minerals are converted into hydrated secondary minerals with an accompanying release of cations and silicic acid:

$$\text{cation Al-silicate} + \text{H (anion)} + H_2O \rightarrow$$

$$\text{Al-silicate} + \text{cation} + H_4SiO_4 + \text{anion}$$

In most cases, dissociated H_2CO_3 (weak carbonic acid) is the H^+ donor in the hydrolysis reaction, for example:

$$2KAlSi_3O_8 \text{ (K-feldspar)} + 2H_2CO_3 + 9H_2O \rightarrow$$

$$Al_2Si_2O_5 (OH)_4 \text{ (kaolinite)} + 2K^+ + 4H_4SiO_4 + 2HCO_3^-$$

This acid derives partly from the solution of carbon dioxide from the atmosphere in precipitation and, more importantly, from the soil atmosphere which may have a CO_2 content up to a few hundred times greater as a result of plant respiration and bacterial activity. The second example relates to the dissolution of carbonate rocks and more particularly limestone, which again occurs widely throughout the world. In this case an acid/base reaction is involved:

$$CaCO_3 + H_2CO_3 \rightarrow Ca^{2+} + 2HCO_3^-$$

or

$$CaMg (CO_3)_2 + 2H_2CO_3 \rightarrow Ca^{2+} + Mg^{2+} + 4HCO_3^-$$

In this instance the weathering reaction is accompanied by a release of cations and the bicarbonate anion.

It is important to note that in both the weathering reactions cited above not all of the soluble products that will be taken up by the water represent rock-derived solutes or a true denudation component. In the first case, the bicarbonate ion is the product of atmospheric or soil-derived CO_2 rather than the rock, and in the second case only 50 per cent of the bicarbonate ion produced can be ascribed to the carbonate minerals involved, the remainder again representing a non-denudational source. Similarly, some of the increase in solute concentration associated with the movement of water through a drainage basin may simply reflect a concentrating effect associated with evapotranspiration losses, analagous to the progressive increase in salinity of a solution that is evaporated to dryness.

Returning to a general global perspective, it is instructive to consider the vast inputs of erosive energy associated with the hydrological cycle. Because of the complex relationship between rainfall intensity, drop size distribution and kinetic energy discussed above, and the importance of snow rather than rain in some areas of the world, it is difficult to estimate the precise level of energy associated with rainfall impact at the land surface of the globe. However, the mean annual precipitation over the land has been estimated at 864 mm (USSR National Committee for the International Hydrological Decade, 1974) and if an average kinetic energy of $20 \, J \, m^{-2} \, mm^{-1}$ of rainfall is assumed from Figure 3.4, the value involved must be of the order of $2 \times 10^{18} \, J \, year^{-1}$. In the case of runoff, an indication of the energy available for erosion and transport can be obtained from an estimate of its *potential* energy. With a total volume of runoff of $47 \times 10^3 \, km^3$ and an average altitude for the land area of the globe (excluding Antarctica) of 700 m, the total potential energy involved is of the order of $3 \times 10^{20} \, J \, year^{-1}$. Although chemical weathering reactions can be studied from the viewpoint of chemical thermodynamics in terms of energy changes and transfers (see, for example, Curtis 1976a and b), this discussion is concerned primarily with the removal of the weathering products and their transport to the oceans. As such it is appropriate to emphasize the annual volume of runoff in contact with the soil and rock of the land surface—that is, $47 \times 10^3 \, km^3$. A $50 \, mg \, l^{-1}$ increase in the solute concentration of this water would represent a total mass of nearly 2.5×10^9 tonnes which is equivalent to a rock volume of approximately $1.0 \, km^3$.

4.3 MATERIAL TRANSPORT TO THE OCEANS

Suspended sediment and dissolved load data are now available for many of the major rivers of the world and for many smaller rivers flowing into the oceans. Taking these data and extrapolating them to areas for which no information is available it is possible to calculate the average annual material transport from the land to the oceans. Rivers draining to inland seas, such as the Caspian and Aral Seas, and other areas of internal drainage are conventionally excluded from these calculations. Furthermore, it must be recognized that the river measurements relate to the lowest gauging site on the river's course, and that a significant proportion of the suspended sediment may be deposited in a delta or estuary before reaching the open sea. Several workers have produced such estimates of material transport to the oceans (Table 4.2), but the validity of

Table 4.2   Some existing estimates of suspended sediment and dissolved load transport to the oceans

Author	Estimated mean annual load (10^9 tonnes)
I *Suspended sediment*	
Fournier (1960)	58.1
Kuenen (1950)	32.5
Gilluly (1955)	31.7
Jansen and Painter (1974)	26.7
Schumm (1963)	20.5
Holeman (1968)	18.3
Goldberg (1976)	18.0
USSR National Committee for the IHD (1974)	15.7
Milliman and Meade (1983)	13.5
Lopatin (1952)	12.7
Mackenzie and Garrels (1966)	8.3
II *Dissolved load*	
Goldberg (1976)	3.9
Livingstone (1963)	3.8
Meybeck (1979)	3.7
Clarke (1924)	3.7
Meybeck (1976)	3.3
Alekin and Brazhnikova (1960)	3.2

their results depends heavily upon the data employed and the degree and method of extrapolation involved. The most reliable estimate of total suspended sediment transport currently available is that provided by Milliman and Meade (1983), whereas the work of Meybeck (1979) represents the most detailed information produced for dissolved loads. The estimate of contemporary suspended sediment transport to the oceans derived by Milliman and Meade (1983) is based on *actual* sediment loads and therefore excludes sediment deposited in major reservoirs which would formerly have reached the oceans. For example, available data suggest that the suspended load at the mouth of the River Nile before the construction of the Aswan High Dam was about 100×10^6 tonnes year^{-1} whereas now it is effectively zero. Similarly Milliman and Meade (1983) suggest that about 30×10^6 tonnes year^{-1} are currently trapped behind the dams on the Zambezi River. Contemporary suspended sediment transport to the oceans in the absence of major dams can therefore be estimated at approximately 14×10^9 tonnes, whilst the total dissolved load amounts to about 4×10^9 tonnes. No reliable data are available for the transport of coarse material as bed load, but if, as many writers have suggested, this can be assumed to amount to about 10 per cent of the suspended load (see, for example, Gregory and Walling, 1973), a tentative estimate of 1×10^9 tonnes can be advanced. The total material transport to the oceans is therefore of the order of 19×10^9 tonnes year^{-1}.

It is not possible to use this value directly to calculate an average rate of lowering of the land surface of the earth. Part of the dissolved load must be attributed to non-denudational sources or to the discharge of man-made pollutants into water courses. According to Meybeck (1979) only 65 per cent of the dissolved load represents the product of chemical denudation and should therefore be included in such calculations. The remaining portion represents atmospheric inputs in precipitation and the uptake of atmospheric CO_2 in weathering reactions. Assuming an average rock density of 2.6 tonnes m^{-3}, these loads denote an average lowering of that portion of the land surface of the earth draining to the oceans (around 105×10^6 km^2) by 64 mm/1000 years. Looking at the relative importance of removal of material as particulates (that is, suspended sediment plus bed load) and in solution, the former accounts for a denudation rate of approximately 55 mm/1000 years and the latter a rate of about 9.5 mm/1000 years, providing a ratio of 6:1 for the relative efficacy of mechanical and chemical denudation processes.

These rates obviously represent averages for the land surface of the globe and, as will be shown later, conceal considerable spatial variation. Furthermore, there are major problems involved in any attempt to derive denudation rates from measurements of sediment load. In the case of suspended sediment and bed load, large quantities of eroded material may be deposited within the drainage basin and may never reach its outlet. River loads will therefore underestimate local rates of surface lowering, but if denudation is viewed as the ultimate *loss* of material from the land to the oceans rather than *redistribution* of material on the land masses, the figures presented above are meaningful. Viewed within the timescale of thousands of years, this movement represents the gradual downwearing of the landscape, although the relatively low rates of lowering involved emphasize the long periods required for landform evolution. At the longer timescale of millions of years, this transport also represents an essential part of the fundamental cycle of erosion, geosynclinal sedimentation and orogeny that has formed the land surface of the earth as we now know it. Sedimentary rocks exposed high in the Alps or other mountain belts may represent sediment deposited in the oceans by rivers of an earlier geological period. An individual particle could conceivably have been recycled several times by the process of deposition, uplift and erosion during the past 600 million years. When viewed in this context, it is of course important to recognize that the present-day sediment loads of rivers may be considerably increased above what might be viewed as 'natural' levels, as a result of the impact of human activity such as forest clearance and cultivation. Equally, moving back through time into the geological past, one can speculate that rates of material transport to the oceans may have been very much higher from earlier land masses. Differences in relief, atmospheric composition and climate must be considered, but more important is the fact that significant colonization of the land surface by vegetation did not occur before the Devonian period (around

400 million years ago) and grasses did not appear until Miocene time (around 25 million years ago). Rates of erosion would have been much greater from bare unvegetated surfaces.

Some information concerning spatial variation of material transport to the oceans across the land surface of the globe may be introduced by comparing, first, the loads transported from the individual continents (Table 4.3) and, secondly, the loads of the world's major rivers. The estimates of the loads transported from the continents listed in Table 4.3 are restricted to suspended sediment and to dissolved material, because of the lack of information on bed load, and are again based on the studies undertaken by Milliman and Meade (1983) and Meybeck (1979). The two sets of load estimates are not strictly equivalent, because the authors employed different data sources and conventions (for example, continental areas) in their derivation. Gross comparisons of the two load components are, nevertheless, possible and the suspended/dissolved load ratios for each continent are listed in Table 4.3. In order to facilitate comparisons of load magnitude between the continents, these data have also been weighted according to continental area and runoff volume.

Looking at the absolute magnitude of the suspended sediment loads for the individual continents listed in Table 4.3, it can be seen that Asia provides the highest load and Europe the lowest. Taking account of the relative size of the continents, however, (see Table 4.3, column 4) maximum specific yields (that is, yields per unit area) are in fact associated with Oceania and the Pacific Islands and minimum values with Africa and Europe, with yields from Oceania and the Pacific Islands exceeding those from Africa by an order of magnitude. Any attempt to account for such contrasts between the continents must consider both the *erosivity* (erosive energy) of the rainfall and runoff regime and the *erodibility* of the terrain. When the specific yields are weighted according to total runoff volume (see Table 4.3, column 5) contrasts between the continents are less apparent, indicating that much of the difference in specific yield can be accounted for in terms of the volumes of runoff involved. However, Oceania and the Pacific Islands again stand out as exhibiting a yield value which is an order of magnitude greater than those for the other continents. This can in turn be related to the highly erodible terrain of many of the large islands of the western Pacific, including Japan, New Guinea, the Philippines, Indonesia, Taiwan and New Zealand, where steep unstable slopes, tectonic activity and intense human activity combine with high rainfall and runoff to produce extremely high sediment yields. Milliman and Meade (1983) have for example demonstrated that the annual suspended sediment yield of the rivers of Taiwan is only slightly less than that from the whole of the mainland United States.

The absolute values of dissolved load for the individual continents listed in Table 4.3 indicate a maximum for Asia and a minimum for Africa. However, the specific yields (see Table 4.3, column 7) provide a somewhat different ranking, with maximum values for Europe, although the overall range of yields

Table 4.3 Material transport from the continents to the oceans based on load estimates produced by Milliman and Meade (1983) and Meybeck (1979)

Continent	1 Land area (10^6 km^2)[3]	2 Mean annual runoff (10^3 km^3)[4]	3 Total annual suspended sediment load (10^6 tonnes year^{-1})[4]	4 Specific suspended sediment yield (tonnes km^{-2} year^{-1})	5 Column 4/ column 2
Africa	15.3	4.1	530	35	8.5
Asia[1]	28.1	14.3	6433	229	16.0
Europe	4.6	2.1	230	50	23.8
North and Central America	17.5	7.8	1462	84	10.8
Oceania/Pacific Islands[2]	5.2	2.4	3062	589	245.4
South America	17.9	11.7	1788	100	8.5

Continent	6 Total annual dissolved load (10^6 tonnes year^{-1})	7 Specific dissolved load (tonnes km^{-2} year^{-1})	8 Column 7/column 2	9 Sediment/dissolved load ratio	10 Overall denudation rate (mm1000 year^{-1})[5]
Africa	201	13	3.2	2.6	17
Asia[1]	1592	57	4.0	4.0	102
Europe	425	92	43.8	0.5	42
North and Central America	758	43	5.5	1.9	43
Oceania/Pacific Islands[2]	293	56	23.3	10.5	241
South America	603	34	2.9	3.0	47

[1] Includes Eurasian Arctic
[2] Includes Australia and the large Pacific Islands
[3] Based on Milliman and Meade (1983)
[4] Based on USSR National Committee for the International Hydrological Decade (1974)
[5] Assuming a 65 per cent denudation component for dissolved load and a rock density of 2.6 tonnes m^{-3}

is much less than for suspended sediment. Runoff volumes should again be considered as an important control, since runoff is responsible for removing and transporting the products of chemical weathering. The specific yields weighted for runoff (see Table 4.3, column 8) evidence little distinction between Africa, the Americas, Africa and Asia, but Europe and to a lesser extent Oceania and the Pacific Islands, are notable for their high values. In the case of Europe the elevated loads can be related to the high proportion of sedimentary strata, which are particularly susceptible to chemical denudation, to the intensive agricultural land use which promotes chemical weathering through soil disturbance and to the effects of pollution from both industrial and agricultural sources in increasing river loads. The high loads indicated for Oceania and the Pacific Islands can be linked to the favourable geological and climatic conditions which exist in many regions of this continent.

Values of the particulate/dissolved load ratio for the individual continents listed in Table 4.3 (column 9) again exhibit considerable variability. In Europe the dissolved load exceeds that of suspended sediment, whilst the suspended sediment load is an order of magnitude greater than the dissolved load for Oceania and the Pacific Islands. As outlined above, it is difficult to interpret these figures directly in terms of the relative importance of chemical and mechanical denudation. However, if, as a gross generalization, it is assumed that about 65 per cent of the dissolved loads represent the product of chemical weathering, then it can be suggested that mechanical and chemical denudation are approximately equal for Europe, whereas mechanical denudation dominates for all other continents and exceeds chemical denudation by a maximum of 16 times in the case of Oceania and the Pacific Islands. Overall rates of denudation for the individual continents based on both suspended sediment and dissolved load, range from a minimum of 17 mm 1000 year^{-1} for Africa to a maximum of 241 mm 1000 year^{-1} for Oceania and the Pacific Islands (Table 4.3, column 10).

Table 4.4 lists current estimates of suspended sediment and dissolved load transport to the oceans by a number of major world rivers. The loads quoted are based on the global data compilations undertaken by Milliman and Meade (1983) and Meybeck (1976, 1984). In several cases they are based on limited measurement programmes and are therefore only approximations. The list includes the eight rivers with the highest suspended sediment loads in rank order and also four rivers with large basin areas but relatively low suspended sediment loads. In looking for the river that transports the maximum suspended sediment load to the oceans, it is questionable whether this distinction should be accorded to the Ganges/Brahmaputra or the Hwang Ho. The former heads the list in Table 4.4, but if it is argued that these two rivers should be treated separately, since they only join in their common delta, then the Hwang Ho assumes this position. There is, however, no question that the Hwang Ho tops the list of world major rivers in terms of both specific sediment yield and when the total

Table 4.4   Suspended sediment and dissolved loads of major world rivers

River	Drainage area (10^6 km^2)	Load (10^6 tonnes) Suspended sediment	Dissolved	Ratio Sediment/ dissolved
High magnitude				
Ganges/Brahmaputra	1.48	1670	151	11.1
Hwang Ho (Yellow)	0.77	1080	34	31.8
Amazon	6.15	900	290	3.1
Yangtze	1.94	478	166	2.9
Irrawaddy	0.43	265	91	2.9
Magdalena	0.24	220	28	7.9
Mississippi	3.27	210	131	1.6
Orinoco	0.99	210	50	4.2
Low magnitude				
Zaire	3.82	43	47	0.91
Ob	2.50	16	50	0.32
Lena	2.50	12	85	0.14
Yenesei	2.58	13	73	0.18

Suspended sediment loads are based on estimates provided by Milliman and Meade (1983), and dissolved loads on estimates compiled by Meybeck (1976, 1984).

load is weighted according to runoff volume and expressed as a mean sediment concentration. Together, the Ganges/Brahmaputra and the Hwang Ho account for almost 20 per cent of the total annual suspended sediment transport to the oceans. The Amazon ranks third on the basis of its total suspended sediment load, but it is notable that its catchment area and runoff volume are far in excess of the combined values for the Ganges/Brahmaputra and the Hwang Ho. This emphasizes the higher rates of mechanical denudation operating in Asian river basins.

Table 4.4 indicates that the Amazon achieves the distinction of transporting the greatest dissolved load of any world river. Values of the suspended sediment/dissolved load ratio for these eight major rivers demonstrate considerable variability. In all cases the particulate load dominates, but the range of values between 11.0 and 1.6 again emphasizes contrasts in the relative efficacy of mechanical and chemical denudation in different river basins.

The values of suspended sediment and dissolved load listed for the four major rivers with low magnitude sediment yields in Table 4.4 highlight the low rates of material transport by rivers draining large areas of tropical Africa and the Eurasian Arctic. In all of these rivers the dissolved load exceeds the particulate load. The low rates of sediment transport by the rivers flowing northward from the Eurasian Arctic may be ascribed to the low-lying terrain which was heavily glaciated during the Pleistocene and therefore provides a low energy environment with limited sources of sediment and to the importance of snow and snowmelt in the overall water budget. The Zaire river is the second largest river in the world in terms of basin area and total runoff volume and in this case the low sediment yield may be related to the subdued topography and dense cover of tropical forest occurring over much of the basin and to the presence of numerous lakes in the lower reaches of the river which act as depositional sinks for transported sediment.

4.4   GLOBAL PATTERNS OF DENUDATION

Further discussion of the contrasts in the magnitude of material transport shown by the individual continents and the major rivers of the world can profitably consider the data available from smaller river basins and for measuring points situated on tributary rivers as well as rivers flowing into the oceans. This change of scales does, however, have important implications for any analysis of suspended sediment loads because it cannot be assumed that all the sediment transported by an upstream tributary will be carried through the river network and therefore be included in the sediment load discharged to the oceans. Much of the sediment may be deposited as it moves through the system and, although it is difficult to generalize, available evidence suggests that only 10 per cent or even less of the suspended sediment load transported from a small (for example, $0.1-1.0 \, km^2$) basin will find its way to the outlet of a large basin (for example, $10\,000 \, km^2$). The many uncertainties associated with any attempt to define this overall process of sediment delivery have been reviewed by Walling (1983). An important distinction must, however, be made between suspended sediment and dissolved load in this context, since the latter is much more conservative in its behaviour and the majority of the dissolved material transported by small tributary streams will be transferred directly to the basin outlet. In any attempt to compare the relative magnitude of the particulate and dissolved components of river load, and therefore the relative importance of mechanical and chemical denudation, it is important to recognize the existence of this scale effect. With other factors remaining constant, the relative importance of the particulate component of a river's load will tend to decline as the size of the river basin increases.

4.4.1   Suspended sediment yields

Measurements of suspended sediment transport are now available for a considerable number of rivers in different parts of the world, and although there are areas for which no data are available, it is possible to introduce a number of general observations concerning both the magnitude of the loads and their spatial variation. Global minima for mean annual specific suspended sediment yield lie well below 2 tonnes km^{-2} year^{-1}. For example, Douglas (1973) cites yields of 1.7 tonnes km^{-2} year^{-1} for the Queanbeyan River (172 km^2) which

Table 4.5   Maximum values of mean annual specific suspended sediment yield for world rivers

Country	River	Drainage area (km^2)	Mean annual suspended sediment yield (tonnes km^{-2} year^{-1})	Source
People's Republic of China	Huangfuchuan	3199	53 500	Yellow River Conservancy Commission (Personal communication)
	Dali	96	25 600	Mou and Meng (1980)
	Dali	187	21 700	Mou and Meng (1980)
Taiwan	Tsengwen	1000	28 000	Milliman and Meade (1983)
Kenya	Perkerra	1310	19 520	Dunne (1975)
Java	Cilutung	600	12 000	Hardjowitjitro (1981)
	Cikeruh	250	11 200	Hardjowitjitro (1981)
New Guinea	Aure	4360	11 126	Pickup, Higgins and Warner (1981)
North Island, New Zealand	Waiapu	1378	19 970	Griffiths (1982)
	Waingaromia	175	17 340	Griffiths (1982)
	Hikuwai	307	13 890	Griffiths (1982)
South Island, New Zealand	Hokitika	352	17 070	Griffiths (1981)
	Cleddau	155	13 300	Griffiths (1981)
	Haast	1020	12 736	Griffiths (1981)

drains the Southern Tablelands and Highlands of New South Wales, Australia, and values of < 1.0 tonnes km^{-2} year^{-1} have been reported for several rivers in Poland (Branski, 1975). Maximum reported mean annual suspended sediment yields exceed 10 000 tonnes km^{-2} year^{-1} and Table 4.5 lists a number of rivers for which such extreme values have been recorded. The highest value contained in Table 4.5 is a mean annual yield of 53 500 tonnes km^{-2} year^{-1} for the Huangfuchuan River (3199 km^2) in the People's Republic of China. This river is a tributary of the middle reaches of the Yellow River which drains the gullied loess region, and the very high suspended load transported by the main Yellow River has already been highlighted in Table 4.4.

Any attempt to account for the extremely high values of sediment yield listed in Table 4.5 must take account of several contributing factors reflecting the erodibility of the terrain and the erosivity of the hydrometeorological regime. In the case of the tributaries of the Yellow River, the existence of highly erodible loess soils, the lack of vegetation cover, and the semiarid climate with intense storm rainfall are major controlling factors (Walling, 1981). The semiarid climate again represents an important causal factor in the Kenyan example, but here severe disturbance of the catchment by overgrazing must also be invoked. For Taiwan, Java and New Guinea, the steep relief, high rainfall totals, and intense agricultural activity are important and in South Island, New Zealand the steep relief and very high rainfall (up to 9000 mm year^{-1}) may again be cited, although tectonic instability also plays a significant role.

Figure 4.4 represents an attempt by the author to produce a generalized global map of suspended sediment yields based on data from more than 1500 measuring stations. The values relate to sediment yields from intermediate-sized basins of the order of 10^4–10^5 km^2. The resultant pattern is of necessity extremely generalized but evidences complex control by a variety of factors, including climate, topography, geology and land use. A number of authors have suggested that sediment yields will be highest in areas of semiarid climate (see, for example, Langbein and Schumm, 1958) and the high yields mapped for the Mediterranean, Southwest United States, and parts of East Africa may be largely ascribed to this tendency. Equally, however, the high sediment yields occurring throughout much of Asia and in the Pacific Islands reflect the high annual rainfall of these areas, although the steep terrain and tectonic instability are also important influences. The close association between high sediment yields and mountain belts is also evident from Figure 4.4, with large areas in the Andes, the Himalayas, Alaska and the Mediterranean producing high yields. The influence of topography and geology is also demonstrated by the low yields mapped for much of the northern regions of Eurasia and North America. Here the subdued relief, widespread glacial deposits and the resistant basement geology are important controls. The extensive areas of low sediment yields in equatorial Africa, and South America are a reflection of the subdued topography and the dense cover of tropical vegetation.

There have been very few attempts to produce a detailed analysis of the various patterns described above in terms of global morphoclimatic zones, but a notable exception is the work of the Soviet scientists Dedkov and Mozzherin published in 1984 (Dedkov and Mozzherin, 1984). These authors assembled sediment load data from more than 3000 measuring stations and firstly subdivided these data into *plains* rivers and *mountain* rivers. Specific suspended sediment yields associated with the latter were on average more than three times greater than those for the former. They subsequently attempted to characterize each of the morphoclimatic zones within these two areas by typical specific suspended sediment yields. Yield values were given for both small (< 5000 km^2) and large (> 5000 km^2) rivers in order to take account of the scale factor discussed above. The results of this analysis presented in Figure 4.5 are heavily dependent on the representativeness of the sediment yield data available for the different morphoclimatic zones because the values depicted for each zone are simply averages of the available data. Figure 4.5 should therefore not be viewed as a definitive representation of the global zonation of sediment yields, but it

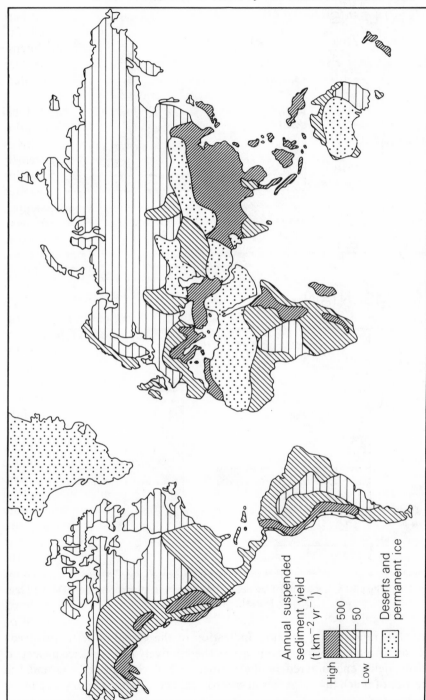

Figure 4.4 A generalized map of the global pattern of specific suspended sediment yield based on data from over 1500 measuring stations

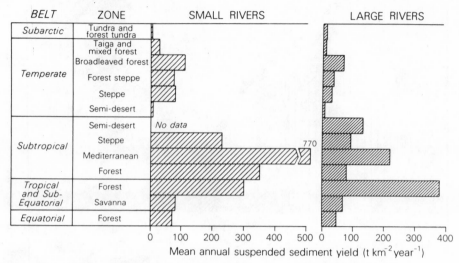

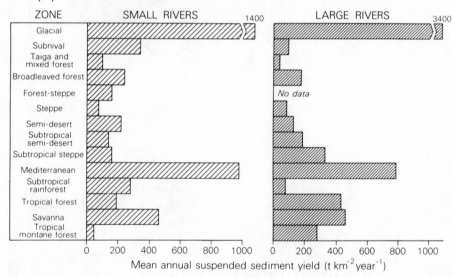

Figure 4.5   The global zonation of suspended sediment yields proposed by Dedkov and Mozzherin (1984)

nevertheless represents a valuable indication of the global patterns involved. In the case of plains rivers, it emphasizes the relatively low specific suspended sediment yields encountered in the temperate and equatorial belts and the occurrence of much higher values in subtropical and tropical regions. A similar

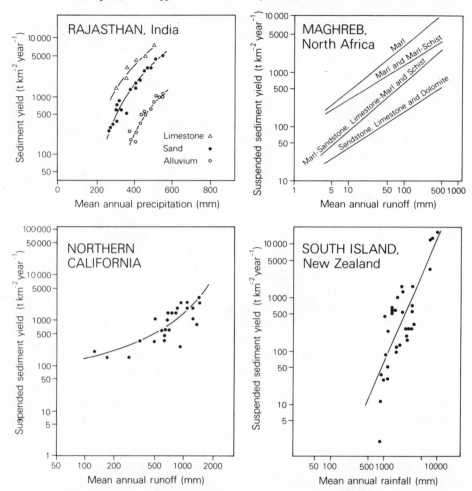

Figure 4.6 Factors controlling regional variations in suspended sediment yield. Based on the work of Sharma and Chatterji (1982) (Rajasthan); Heusch and Milliès-Lacroix (1971) (Maghreb); Janda and Nolan (1979) (Northern California); and Griffiths (1981) South Island, New Zealand

pattern is evident for mountain rivers, but in this case the rivers in the glacial zones produce the highest levels of sediment yield.

In considering the generalized information on global patterns presented in Figures 4.4 and 4.5, it is important to appreciate that very considerable local variation in suspended sediment yield may occur within a particular region or zone in response to a variety of controlling factors. Examples of results from several studies which have attempted to analyse and account for these local variations are presented in Figure 4.6. In each case the magnitude of mean annual

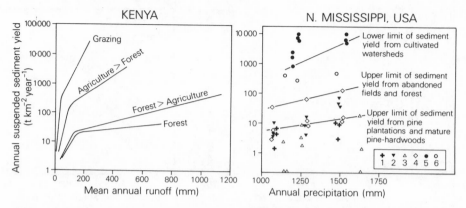

Figure 4.7   The influence of land use on suspended sediment yields. The example from Kenya is based on the work of Dunne (1979), and that from northern Mississippi on the work of Ursic and Dendy (1965). In the latter example the symbols refer to watershed land use: 1 : mature pine-hardwoods, 2 : depleted hardwoods; 3 : pine plantations; 4 : abandoned fields; 5 : cultivated; 6 : pasture

precipitation or mean annual runoff has been shown to exert a major control on sediment yield. These positive relationships between sediment yield and annual precipitation and runoff can be interpreted as a response to the erosivity of the hydrometeorological regime. In the studies undertaken in Rajasthan and in the Maghreb region of North Africa, additional variation was accounted for by consideration of the local soil and geology, although a simple geological variable will inevitably subsume many other catchment characteristics, including topography, vegetation and land use, which generally reflect the erodibility of the terrain. In these two areas, local variations in erosivity and erodibility can be seen to have approximately equal weight in accounting for the observed range of sediment yield.

Land use will exert a major influence on the magnitude of sediment yield from a drainage basin since it will influence the degree of protection afforded by any vegetation cover, the physical properties of the soil and the potential for surface runoff generation. There have been many studies that have demonstrated substantial contrasts in erosion and sediment yield as a result of differences in land use, and two are illustrated in Figure 4.7. In the example from Kenya based on the work of Dunne (1979) a positive relationship was developed between mean annual runoff and sediment yield, but the precise form of the relationship varied according to the land use. On the basis of these relationships it can be hypothesized that a change in land use from native forest to intensive grazing could produce an increase in sediment yield of as much as three orders of magnitude. Similarly, in the example drawn from the work of Ursic and Dendy (1965) in northern Mississippi, United States, a positive relationship between mean annual precipitation and sediment yield was shown

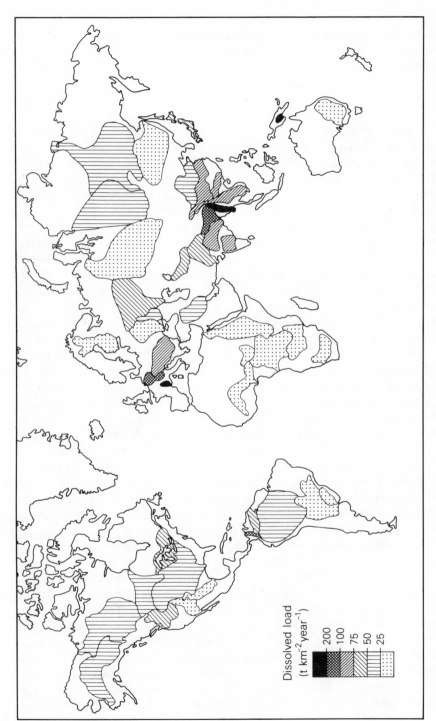

Figure 4.8   The specific yield of dissolved material from the major river basins of the world. Based on data presented by Meybeck (1976, 1984) and other authors

Dissolved load
(t km^{-2} year^{-1})

200
100
75
50
25

to be clearly differentiated according to land use, with sediment yields from cultivated watersheds exceeding those from forest areas by two to three orders of magnitude.

4.4.2 Dissolved loads

In comparison to suspended sediment loads, much less information is available on the dissolved loads of rivers. However, a general review of existing data produced by Walling and Webb (1986) and which assembled information from nearly 500 rivers in different areas of the world provides a useful assessment of the range of values likely to be encountered. These authors reported minimum values of less than 1.0 tonnes km^{-2} year^{-1} for three intermediate-sized drainage basins in southern Alberta, Canada, a maximum value of 500 tonnes km^{-2} year^{-1} for the River Dranse in the Chablais region of France, and a mean value for the data set of 38.8 tonnes km^{-2} year^{-1}. Other authors have since cited estimates of higher loads, and Meybeck (1984) refers to values of 6000 tonnes km^{-2} year^{-1} for the River Cana in Amazonia which drains an area of halite deposits, and of 750 tonnes km^{-2} year^{-1} for an area of karst in Papua New Guinea. The first of these values must be viewed as a local anomaly, but the other can be accepted as providing a realistic indication of the global maximum value. As such, dissolved loads span a considerably smaller range than suspended sediment loads.

In the absence of dissolved load data for many areas of the world, it is impossible to produce a generalized map of solute yields which is comparable to Figure 4.4. However, in Figure 4.8 the available data for major world rivers compiled by Meybeck (1976, 1984) and a number of other authors have been mapped. In general, the resulting pattern is somewhat easier to explain than that for suspended sediment yields and can be interpreted as largely reflecting the interaction of runoff amount and lithology and to a lesser extent the temperature regime. The high values of load mapped for many Asian rivers are a response to their high runoff totals which provide greater opportunity for removal and transport of solutes from their drainage basins. The relatively high loads indicated for several European rivers can equally be linked to the predominance of sedimentary rocks, including limestones, within their catchments. Conversely the low dissolved loads encountered in Africa and Australia reflect the existence of ancient basement rocks with a low susceptibility to chemical weathering. The extremely high load values for the Irrawaddy and for the Fly and Putari rivers in Papua New Guinea undoubtedly reflect both high runoff totals and the presence of readily weathered sedimentary strata, but the high temperatures associated with their tropical climates may also be important in promoting rapid chemical weathering.

Further evidence of the influence of runoff magnitude, lithology and temperature regime in controlling global patterns of dissolved load is provided in Figure 4.9. Figures 4.9a and 4.9b present relationships between both discharge-weighted total dissolved solids concentration and mean annual dissolved load,

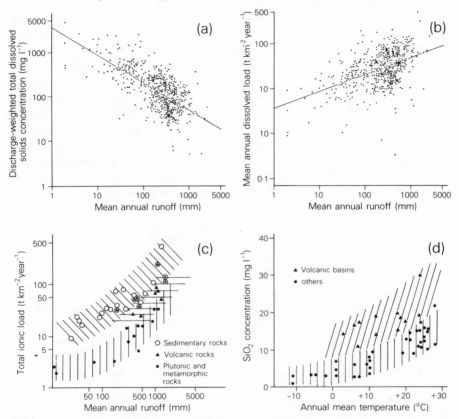

Figure 4.9   Factors controlling global variations in dissolved load transport. (a),(b) Based on Walling and Webb (1986); (c),(d) based on Meybeck (1981)

and mean annual runoff for the global database of 496 rivers compiled by Walling and Webb (1986). In the case of the plot of concentration versus mean annual runoff, the inverse relationship may be accounted for in terms of a general dilution effect, as runoff volumes increase. The clear positive relationship between load and annual runoff nevertheless indicates that increasing moisture availability provides an increase in the total amount of dissolved material released or available for transport and emphasizes the importance of runoff magnitude in controlling the global pattern of loads. Much of the scatter evident in Figure 4.9b can be ascribed to the effects of lithology in influencing rates of chemical weathering and solute release. Meybeck (1981) has attempted to demonstrate this influence by further subdividing a similar relationship between total ionic load and annual runoff according to rock type and associated susceptibility to chemical denudation (Figure 4.9c). According to Figure 4.9c, the total ionic load for a given level of annual runoff will be greatest from a basin underlain by sedimentary rocks, and least from an area of plutonic and metamorphic rocks. Volcanic rocks provide intermediate load values. Meybeck (1981) has also effectively demonstrated the influence of temperature regime on the rate of chemical weathering, and therefore dissolved loads, by establishing clear positive relationships between annual mean temperature and the dissolved silica concentration in river water (Figure 4.9d). Again, however, the influence of rock type is evident, since silica concentrations in areas underlain by volcanic rocks are commonly double those found elsewhere (Figure 4.9d).

As in the case of suspended sediment yields, the broad global trends outlined above conceal considerable local variations in dissolved load. Geological controls frequently become increasingly important as the scale of attention reduces. For example, Walling and Webb (1986) report an analysis of countrywide variations in dissolved loads for Great Britain. Here values range between 10 and in excess of 200 tonnes km^{-2} $year^{-1}$ and the national pattern was shown to be controlled primarily by lithology rather than the magnitude of annual runoff.

4.5   WIDER IMPLICATIONS

The discussion of the erosional effects of the global hydrological cycle on the land surface of the earth and the resultant transport of material to the oceans provided above has focused primarily on geomorphological considerations of land denudation. The significance of this material transfer for the long-term geological or exogenic cycle of erosion, geosynclinal sedimentation and orogeny has also been introduced. Adopting a wider perspective, however, it is important to appreciate that the transport of particulate and dissolved material from the continents to the oceans also represents an important pathway in many global geochemical cycles (see, for example, Martin and Meybeck, 1979). For example, the transport of dissolved carbonates and particulate and dissolved organic matter by rivers provides one of the major links in the global carbon cycle (see, for example, SCOPE, 1979). The need to produce an improved assessment of the fluxes associated with this pathway has stimulated a growing interest in the investigation of river loads (see, for example, Likens, 1981). Furthermore, concern for pollution of the oceans has also highlighted the need to document river inputs and the role of runoff in transferring contaminants from the terrestrial to the marine environment (see, for example, Waldichuk, 1977).

References

Alekin, O. A. and Brazhnikova, L. V. (1960). A contribution on runoff of dissolved substances on the world's continental surface. *Gidrochim. Mat.* 32, 12–24.
Arnoldus, H. M. J. (1977). Methodology used to determine the maximum potential average annual soil loss due to sheet and rill erosion in Morocco. In *Assessing Soil Degradation*, FAO Soils Bulletin no. 34, 39–42.

Branski, J. (1975). Ocena denudacji dorzecza Wisley na podstawie wynikow pomiarow rumowiska unoszonego. *Prace Instytutu Meteorologii i Gospodarki Wodnej*, **6**, 1–58.

Bricker, O. P. and Garrels, R. M. (1967). Mineralogic factors in natural water equilibria. In Faust, S. D. and Hunter, J. V. (eds) *Principles and Applications of Water Chemistry*, Wiley: New York, 449–469.

Carter, C. E., Greer, J. D., Braud, H. J., and Floyd, J. M. (1974). Raindrop characteristics in South Central United States. *Transactions American Society of Agricultural Engineers*, **17**, 1033–1037.

Ciccaci, S., Fredi, P., and Lupia Palmieri, E. (1977). Rapporti fra transporto solido e parametri climatici e geomorfici in alcuna basini idrografici Italiani. In *Misura del Transporto Solido al Fondo Nei Corsi d'Acqua: Problemi per Una Modellistica Matematica*. Firenze: Instituto de Ingegneria Civile, C4.1–C4.16.

Clarke, F. A. (1924). The data of geochemistry. *US Geological Survey Bulletin* **770**.

Curtis, C. D. (1976a). Chemistry of rock weathering: fundamental reactions and controls. In Derbyshire, E. (ed.) *Geomorphology and Climate*. Wiley: Chichester, 25–57.

Curtis, C. D. (1976b). Stability of minerals in surface weathering reactions: a general thermochemical approach. *Earth Surface Processes*, **1**, 63–70.

Dedkov, A. P. and Mozzherin, V. I. (1984). *Eroziya i Stok Nanosov na Zemle*, Izdatelstvo Kazanskogo Universiteta, 264 pp.

Demmak, A. (1982). *Contribution a l'étude de l'érosion et les transports solides en Algérie septentrionale*. Thèse de Docteur-Ingenieur, Université de Paris.

Douglas, I. (1973). *Rates of Denudation in Selected Small Catchments in Eastern Australia*. University of Hull, Occasional Papers in Geography, no. 21, 127 pp.

Dunne, T. (1975). Sediment yields of Kenyan rivers. Unpublished report.

Dunne, T. (1979). Sediment yield and land use in tropical catchments. *Journal of Hydrology*, **42**, 281–300.

Fournier, F. (1960). *Climat et Erosion*. Presses Universitaires de France: Paris, 201 pp.

Gilluly, J. (1955). Geologic contrasts between continents and ocean basins. *Geological Society of America Special Paper*, **62**, 7–18.

Goldberg, E. D. (1976). *The Health of the Oceans*. UNESCO: Paris, 172 pp.

Gregory, K. J. and Walling, D. E. (1973). *Drainage Basin Form and Process*. Edward Arnold: London, 458 pp.

Griffiths, G. A. (1981). Some suspended sediment yields from South Island catchments, New Zealand. *Water Resources Bulletin*, **17**, 662–671.

Griffiths, G. A. (1982). Spatial and temporal variability in suspended sediment yields of North Island basins, New Zealand. *Water Resources Bulletin*, **18**, 575–584.

Hardjowitjitro, H. (1981). Soil erosion as a result of traditional cultivation in Java Island. In Tingsanchali, T. and Eggers, H. (eds), *Proceedings of the South-east Asian Regional Symposium on Problems of Soil Erosion and Sedimentation*. Asian Institute of Technology: Bangkok, 173–179.

Heusch, B. and Milliès-Lacroix, A. (1971). Une méthode pour estimer l'écoulement et l'érosion dans un bassin. Application au Maghreb. *Mines et Geologie* (Rabat) no. 33.

Holeman, J. N. (1968). The sediment yield of major rivers of the world. *Water Resources Research*, **4**, 737–747.

Janda, R. J. and Nolan, K. M. (1979). Stream sediment discharge in Northwestern California. In *Guidebook for a Field Trip to Observe Erosion in Franciscan Terrane of Northern California*, US Geological Survey: Menlo Park, California, IV-1–IV-27.

Jansen, J. H. L. and Painter, R. B. (1974). Predicting sediment yield from climate and topography. *Journal of Hydrology*, **21**, 371–380.

Kuenen, P. H. (1950). *Marine Geology*, Wiley and Sons: New York, 568 pp.

Lal, R. (1976). *Soil Erosion Problems on an Alfisol in Western Nigeria and Their Control.* International Institute of Tropical Agriculture Monograph no. 1: Ibadan, Nigeria.

Langbein, W. B. and Schumm, S. A. (1958). Yield of sediment in relation to mean annual precipitation. *Transactions of the American Geophysical Union*, **39**, 1076–1084.

Likens, G. E. (ed.) (1981). *Flux of Organic Carbon by Rivers to the Oceans.* US Department of Energy: Washington, DC, 397 pp.

Livingstone, D. A. (1963). Chemical composition of rivers and lakes. Data of Geochemistry, Chapter G. *US Geological Survey Professional Paper* 440G.

Lopatin, G. C. (1952). Detritus in the rivers of the USSR. *Zap. Vses. Geogr. Obsch.* **14**, Geografgiz: Moscow.

Mackenzie, F. T. and Garrels, R. M. (1966). Chemical mass balance between rivers and oceans. *American Journal of Science*, **264**, 507–525.

Martin, J. M. and Meybeck, M. (1979). Elemental mass-balance of material carried by world major rivers. *Marine Chemistry*, 7, 173–206.

Meybeck, M. (1976). Total dissolved transport by world major rivers. *Hydrological Sciences Bulletin*, **21**, 265–284.

Meybeck, M. (1979). Concentrations des eaux fluviales en éléments majeurs et apports en solution aux oceans. *Revue de Géologie Dynamique et de Géographie Physique*, **21**, 215–246.

Meybeck, M. (1981). Pathways of major elements from land to ocean through rivers. In *River Inputs to Ocean Systems*, UNEP/UNESCO, 18–30.

Meybeck, M. (1984). *Les fleuves et le cycle géochimique des éléments.* Thèse d'Etat, Université Pierre et Marie Curie, Paris.

Meyer, L. D. and Wischmeier, W. H. (1969). Mathematical simulation of the process of soil erosion by water. *Transactions American Society of Agricultural Engineers*, **12**, 754–758, 762.

Mihara, Y. (1959). *Raindrops and Soil Erosion.* National Institution of Agricultural Science: Tokyo, Bulletin no. 1, 59 pp.

Milliman, J. D. and Meade, R. H. (1983). World-wide delivery of river sediment to the oceans. *Journal of Geology*, **91**, 1–21.

Mou, J. and Meng, Q. (1980). *Sediment Delivery Ratio as used in the Computation of the Watershed Sediment Yield.* Beijing, China.

Pickup, G., Higgins, R. J., and Warner, R. H. (1981). Erosion and sediment yield in Fly River drainage basins. In Davies, T. R. H. and Pearce, A. J. (eds), *Erosion and Sediment Transport in Pacific Rim Steeplands*, International Association of Hydrological Sciences Publication no. 132, 438–456.

SCOPE (1979). *The Global Carbon Cycle.* (eds B. Bolin, E. T. Degens, S. Kempe and P. Ketner). SCOPE Report 13, Chichester: John Wiley and Sons.

Schumm, S. A. (1963). The disparity between present rates of denudation and orogeny. *US Geological Survey Professional Paper* 454H.

Sharma, K. D. and Chatterji, P. C. (1982). Sedimentation in Nadis in the Indian arid zone. *Hydrological Sciences Journal*, **27**, 345–352.

Stocking, M. A. and Elwell, H. A. (1976). Rainfall erosivity over Rhodesia. *Transactions of the Institute of British Geographers*, **1**, 231–245.

UNESCO (1978). *World Water Balance and Water Resources of the Earth.* UNESCO Studies and Reports in Hydrology no. 25, 663 pp.

Ursic, S. J. and Dendy, F. E. (1965). Sediment yields from small watersheds under various land uses and forest covers. *Proceedings Federal Inter-Agency Sedimentation Conference, US Department of Agriculture Miscellaneous Publication*, **970**, 47–52.

USSR National Committee for the International Hydrological Decade (1974). *Mirovoi Vodnyi Balans i Vodnye Resursy Zemli.* Gidrometeoizdat, Leningrad, 638 pp. (English translation as UNESCO, 1978).

Waldichuk, M. (1977). *Global Marine Pollution: An Overview*. Intergovernmental Oceanographic Commission Technical Series no. 18, UNESCO, Paris, 96 pp.

Walling, D. E. (1981). The Yellow River which never runs clear. *Geographical Magazine*, **53**, 568–575.

Walling, D. E. (1983). The sediment delivery problem. *Journal of Hydrology*, **69**, 209–237.

Walling, D. E. and Webb, B. W. (1986). Solutes in rivers systems. In Trudgill, S. T. (ed.) *Solute Processes*. Chichester: John Wiley and Sons, 251–327.

Wischmeier, W. H. and Smith, D. D. (1978). *Predicting Rainfall-erosion Losses*, Agriculture Handbook no. 537. Washington, DC: US Department of Agriculture.

Energetics of Physical Environment
Edited by K. J. Gregory
©1987 John Wiley & Sons Ltd

5

Energetics of Soil Processes

SHEILA M. ROSS

Department of Geography, University of Bristol

5.1 INTRODUCTION

Energy is universally defined in physics as the capacity to do work. In the soil, energy is the driving force for the pedological processes which control the formation and subsequent development of distinctive soil profiles. Three different sources of energy are available for 'work' in carrying out soil processes (Table 5.1). Chemical energy controls the *formative soil processes* of rock weathering and organic matter decomposition which produce soil material. Mechanical energy controls the *dynamic soil processes* which move mineral particles, soil water and solutes around the soil profile. Solar radiation plays a dual role in soil energetics. Firstly, radiant energy is 'stored' in the form of chemical energy, initially in plant tissues and subsequently in vegetation litter and soil organic matter. Secondly, solar radiation, in the form of heat energy, is a major control on rates of formative chemical processes. Heat and mechanical energy, in the form of gravitational energy, are the main regulators of rates of dynamic, physical processes (Figure 5.1). The very wide range in relative proportions of these available energy forms worldwide result in an extremely large variety of soil profile types. In England and Wales alone, 296 different soil association profile types have been mapped at the 1:250 000 scale while the number of different soil series recognized in the United States now exceeds 10 500.

Energetics or energy transformations involved in soil processes must obey the laws of thermodynamics. The first law of thermodynamics states that the total amount of energy in a system remains constant; energy can be transformed but is neither created nor destroyed. This is the principle of energy conservation and holds true only for systems in which there is no change of mass. An example of the operation of this law is that in soil reactions involving changes in chemical energy, any apparent energy loss usually indicates transformation into heat. Soil reactions occurring without inputs of energy from outside the system and

119

Table 5.1 Sources of energy for soil processes

Form of energy	Description	Energy use in soil processes
Solar radiation	Electromagnetic energy absorbed at the soil surface and (a) transformed into *heat energy* (b) transformed into *chemical energy* in organic molecules of plants by photosynthesis. Their litter transfers chemical energy to soil organic matter	Heat energy warms up soil components (minerals, organic matter, air, water) according to their specific heats. Thermal conductivity and specific heat are used to determine rates of thermal diffusion through soils containing different proportions of minerals, organic matter, air and water.
Mechanical energy	Consists of potential energy (*PE*) and kinetic energy (*KE*). PE = stored energy a body has due to its position in a force field. In soil this is gravity. *PE* is converted to *KE* when a body moves. KE = energy of motion: the amount of work done to bring a body to rest	Controls *dynamic* soil processes. Mechanical energy is used in the translocation of soil water and soil solutes through the soil profile. Examples of such processes include: eluviation/illuviation during podzolization; leaching; lessivage; erosion.
Chemical energy	Stored energy in the chemical bonds of all molecules in soil materials	Controls *formative* soil processes. Chemical energy in bonds of organic and inorganic molecules powers the processes of organic matter decomposition and rock weathering respectively. Changes of state of soil water (for example during evaporation or freezing) involve the transformation of chemical energy into *latent heat*. Respiration of the living soil population involves losses of *heat* energy.

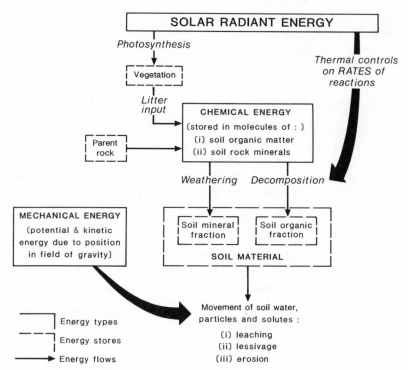

Figure 5.1 Role of solar, chemical and mechanical energy in soil formation and development processes

emitting energy as a byproduct, are *exoergic* reactions (*exothermic* for reactions generating heat energy). Reactions which absorb energy cannot proceed without imports of energy from another source and are termed *endoergic* reactions (*endothermic* if heat energy is absorbed). The degree to which a given amount of work is dissipated as heat instead of being stored as potential energy (*PE*) or kinetic energy (*KE*) is a measure of the irreversibility of the process. Since heat energy is constantly being lost through the decompositional activities and respiration of soil organisms, these biochemical soil processes are highly irreversible.

The second law of thermodynamics states that systems develop through time towards a state of maximum entropy or disorder, with lowest available energy for carrying out work. Young sediments and newly forming soils are at a state of maximum order since they have very simple mineralogies and textures. They also have maximum available energy to perform work. Through time and profile development, soil materials become more disordered as a heterogeneous mixture of mineral and organic compounds are produced during weathering and decomposition. Well-developed, mature soil profiles with very complex physics and chemistries contain minimum amounts of available energy for further work. In the following sections we shall examine the operation of these simple thermodynamic principles in relation to the processes which produce soil material and to translocation processes in developing soil profiles.

5.2 ENERGETICS OF FORMATIVE SOIL PROCESSES

Mineral weathering and organic matter decomposition are the two processes that control the production of soil material. The major difference in energy budgeting for the mineral and organic fractions in the soil relates to the renewability or continued supply of chemical energy. It is useful for thermodynamic purposes to consider the soil mineral fraction as a finite chemical energy resource, with a very slow input of new, weatherable parent material to the system. This allows the mineral part of the soil to be considered in the short to medium term as a closed, but perhaps slightly leaky system. On the other hand, let us consider that the organic fraction in the soil is not finite because additions to this resource through litter fall, root death, animal and organism faeces and their dead body tissues, all occur over short time-scales.

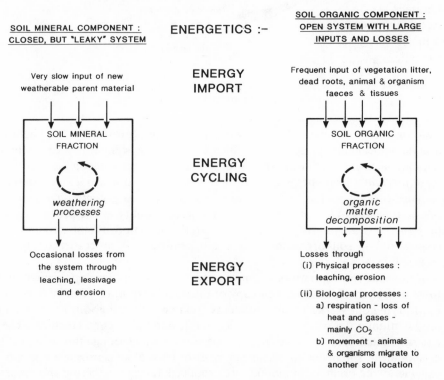

Figure 5.2 Diagrammatic energy budgets for soil mineral and organic fractions

The organic part of the soil can thus be considered as an open system. These ideas are illustrated in Figure 5.2. The main significance of this for energy cycling is that losses of chemical energy from the mineral fraction by physical processes such as erosion may take a long time to be replaced by weatherable parent material. In the case of the organic fraction, the large losses of degraded thermal energy through biological processes, mainly organism movement and respiration, are replaced by inputs of chemical energy in litter and organic matter.

Energy is stored in the chemical bonds of inorganic and organic crystals and molecules. This energy provides the power for the breakdown of rocks through chemical weathering reactions and the biological decomposition of soil organic matter and vegetation litter through microbial activity and respiration. All matter contains chemical energy resulting from (a) the *KE* of its component molecules; (b) the *KE* and *PE* associated with electrons and nuclei in component atoms; and (c) the *PE* associated with the bonds between atoms when they form molecules. This internal energy is the thermodynamic potential of the compound or series of compounds taking part in any chemical reaction. When temperature and pressure are held constant, as is the case in many simple chemical thermodynamic calculations, this internal energy is called the Gibbs free energy (ΔG) of the system. During soil chemical reactions, the breaking and reformation of molecular bonds involves energy transfers. In simplified soil reactions, these energetics can be calculated according to chemical thermodynamics.

5.2.1 Energetics of mineral weathering in soil

The physical break-up of rocks and mineral particles is primarily due to expansion and contraction during wetting/drying, freezing/thawing and crystal growth. The changes in temperature, pressure and soil moisture which control these processes are caused by variations in thermal energy in the soil resulting from differential receipt and transmission of solar radiation. In this way solar energy acts as a rate control on the production of disintegrated mineral material and in increasing the surface area of mineral particles to be further acted on by chemical weathering processes.

In developing soils, the mineral fraction is complex in its composition. Chemical weathering reactions are usually simplified into five reaction types: solution, carbonation, hydrolysis, hydration and oxidation/reduction, to allow the theoretical isolation of discrete and simple chemical reactions. In reality, multistage reactions between reactants and products occur to complicate this picture. For this reason, the theoretical thermodynamic examples discussed below may be considered as being too simple to apply to real soil weathering conditions. Nevertheless, simple chemical thermodynamics are a necessary precursor to any comprehensive synthesis of weathering chemistry and have aided understanding of mineral stability and instability in the soil. Goldich's (1938) classic weathering sequence indicates that silicate minerals occur in the following sequence, from the least to the most resistant:

olivine < augite < hornblende < biotite < potassium feldspar < muscovite < quartz.

Among the feldspars, calcium feldspars are much less stable than sodium feldspars. Curtis (1976) has shown that chemical thermodynamics can account for this sequence, by examining the changes in internal energy that occur when mineral weathering reaction are formulated for realistic soil environmental conditions.

The likelihood of a weathering reaction taking place, or the order in which a series of weathering reactions will occur, can be predicted by comparing the values of Gibbs free energy of formation (ΔG_f° in kJ mol^{-1}) for reactants and products, to obtain a Gibbs free energy for the reaction (ΔG_r°). The superscript ΔG° is used for standard states (solid, liquid or gas) at standard temperature and pressure (25 °C and 100 kPa). It is important to note that chemical thermodynamics tells us nothing about the relative rates of reactions. If we examine a very simple weathering reaction:

$$a_{(solid)} \quad + \quad H_2O_{(liquid)} \quad \rightleftharpoons \quad b_{(solid)} \quad + \quad c_{(aqueous)} \qquad (1)$$

reactant	reactant	product	product
mineral	water	mineral	soluble

then ΔG_r° for the reaction can be calculated as:

$$\Delta G_r^\circ = [\Delta G_f^\circ(b) + \Delta G_f^\circ(c)] \quad - \quad [\Delta G_f^\circ(a) + \Delta G_f^\circ(H_2O)] \qquad (2)$$

Change in Gibbs free energy for the reaction	Sum of Gibbs free energies for all the *products* in their standard states	Sum of Gibbs free energies for all the *reactants* in their standard states

If the ΔG_r° value in Equation (2) is negative, then ΔG_f° (reactants) is greater than ΔG_f° (products) and energy is liberated during the reaction. This means that the products are more stable than the reactants. The reactant mineral, *a*, is unstable under the prevailing environmental conditions, and reaction (1) is likely to take place in the forward direction.

An example of the use of ΔG_r° calculations in studying mineral weathering reactions is given below for the alteration in soil of two primary feldspar minerals. Anorthite, a calcium plagioclase feldspar, and albite, a sodium feldspar, weather eventually to kaolinite, a secondary aluminosilicate clay mineral. These reactions have traditionally been written as simple hydration equations.

Table 5.2 Thermodynamic calculations of ΔG_r° for anorthite and albite weathering reactions, with the production of kaolinite (all ΔG_f° values in kJ mol^{-1} at 25 °C and 100 kPa, calculated from Drever (1982)

(a) Traditionally formulated *hydration* weathering reactions

ANORTHITE

Weathering reaction:
$$CaAl_2Si_2O_8 + 3H_2O \rightarrow Al_2Si_2O_5(OH)_4 + Ca^{2+} + 2OH^-$$
anorthite   water   kaolinite

Free energy:
$$\Delta G_r^\circ = \Sigma\Delta G_f^\circ \text{ (kaolinite} + Ca^{2+} + 2(OH^-)) - \Sigma\Delta G_f^\circ \text{ (anorthite} + 3H_2O)$$
$$= [-3789.01 + (-553.54) + 2(-157.32)] - [-3992.79 + 3(-237.19)]$$
$$= (-4657.25) - (-4704.36)$$
$$\Delta G_r^\circ = +47.11 \text{ kJ mol}^{-1}$$

ALBITE

Weathering reaction:
$$2NaAlSi_3O_8 + 3H_2O \rightarrow Al_3Si_2O_5(OH)_4 + 4SiO_2 + 2Na^+ + 2OH^-$$
albite   water   kaolinite   silica
(quartz)

Free energy:
$$\Delta G_r^\circ = \Sigma\Delta G_f^\circ \text{ (kaolinite} + 4 \text{ quartz} + 2Na^+ + 2OH^-) - \Sigma\Delta G_f^\circ \text{ (albite} + 2H_2O)$$
$$= [-3789.07 + 4(-848.89) + 2(-261.88) + 2(-157.32)] - [2(-3708.32) + 3(-237.19)]$$
$$= (-8023.03) - (-8128.21)$$
$$\Delta G_r^\circ = 105.19 \text{ kJ mol}^{-1}$$

(b) Incorporating anionic (acidic) influence in hydration reactions

ANORTHITE

Weathering reaction:
$$CaAl_2Si_2O_8 + 3H_2O + CO_{2(aqueous)} \rightarrow Al_2Si_2O_5(OH)_4 + Ca^{2+} + 2HCO_3^-$$
anorthite   water   carbon   kaolinite   bicarbonate
dioxide

Free energy:
$$\Delta G_r^\circ = \Sigma\Delta G_f^\circ \text{ (kaolinite} + Ca^{2+} + 2HCO_3^-) - \Sigma\Delta G_f^\circ \text{ (anorthite} + 3H_2O + CO_2)$$
$$= [-3789.07 + (-553.54) + 2(-586.85)] - [-3992.79 + 3(-237.19) + (-394.34)]$$
$$= (-5516.31) - (-5098.71)$$
$$\Delta G_r^\circ = -417.61 \text{ kJ mol}^{-1}$$

ALBITE

Weathering reaction:
$$2NaAlSi_3O_8 + 3H_2O + CO_2 \rightarrow 4SiO_2 + Al_2Si_2O_5(OH)_4 + 2Na^+ + 2HCO_3^-$$
albite   water   carbon   quartz   kaolinite   bicarbonate
dioxide

Free energy:
$$\Delta G_r^\circ = \Sigma\Delta G_f^\circ \text{ (kaolinite} + 4 \text{ quartz} + 2Na^+ + 2HCO_3^-) - \Sigma\Delta G_f^\circ \text{ (2 albite} + 3 \text{ water} + CO_2)$$
$$\Delta G_r^\circ = [4(-856.26) + (-3789.07) + 2(-261.88) + 2(-586.85)] - [2(-3708.32) + 3(-157.32) + (-394.34)$$
$$= (-8911.54) - (-8282.94)$$
$$\Delta G_r^\circ = -628.60 \text{ kJ mol}^{-1}$$

(A) *The weathering of anorthite to kaolinite*

$$CaAl_2Si_2O_{8(solid)} + 3\,H_2O_{(liquid)} \rightarrow Al_2Si_2O_5(OH)_{4(solid)} + Ca^{2+}_{(aqueous)} + 2\,OH^-_{(aqueous)}$$
$$\underset{\text{anorthite}}{} \qquad\qquad\qquad \underset{\text{kaolinite}}{} \qquad\qquad\qquad\qquad\qquad\qquad (3)$$

(B) *The weathering of albite to kaolinite*

$$2NaAlSi_3O_{8(solid)} + 3H_2O_{(liquid)} \rightarrow Al_2Si_2O_5(OH)_{4(solid)} +$$
$$\underset{\text{albite}}{} \qquad\qquad\qquad\qquad \underset{\text{kaolinite}}{} \qquad\qquad\qquad (4)$$

$$4SiO_{2(solid)} + 2Na^+_{(aqueous)} + 2OH^-_{(aqueous)}$$
$$\underset{\text{quartz}}{}$$

In a mixed soil mineral substrate containing anorthite and albite, which of these two minerals will be the first to weather to kaolinite? To answer this question, we must calculate and compare the ΔG_r° values for Equations (3) and (4) above. These calculations are outlined in Table 5.2(a). Both ΔG_r° values are positive. This means that neither reaction will occur spontaneously at standard temperature and pressure. Both weathering mechanisms require environmental 'catalysts' before they can proceed. Examples of such catalysts in soils may be the presence of water, other chemicals in the soil solution, acidic pH, anaerobic conditions or elevated temperatures. These environmental catalysts act as an energy input which allows the system or reaction to overcome any initial energy hurdle, called the activation energy (Figure 5.3), such that:

$$\Delta G_r = E_{af} - E_{ar} \qquad\qquad\qquad (5)$$

where

$E_{af} =$ activation energy for the forward reaction
$E_{ar} =$ activation energy for the reverse reaction
$\Delta G_r =$ net internal energy change for the reaction

Where E_{ar} is substantially greater than E_{af}, the forward reaction is very much more likely to occur than the reverse reaction since a much larger energy input is required for the reverse reaction. So reactions with large E_{ar} values are virtually irreversible.

Since the positive ΔG_r° for anorthite weathering < the positive ΔG_r° for albite weathering, we might conclude that under uniform environmental conditions, the production of kaolinite from albite decay will occur preferentially to kaolinite production from the decay of anorthite. This is because more energy would have to be added to the weathering system for equation (4) to occur than for equation (3) to occur. Both reactions appear to produce alkaline soil solutions through the release of OH^- ions during hydration. Since soil solutions and drainage waters are much more commonly neutral to acidic, the relevance of

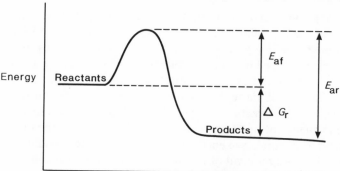

Figure 5.3 Energy levels of reactants and products in a chemical reaction, illustrating the activation energy for the forward and reverse reactions (ΔG_r, E_{af} and E_{ar} defined as in Equation (5))

such simple calculations for weathering reactions remains doubtful. Curtis (1975) suggests that it would be more realistic to formulate weathering equations to include the influence of acid inputs in rainfall. Carbonic acid formed by the dissolution of atmospheric carbon dioxide in rain water, and dilute sulphuric acid formed from the dissolution of sulphurous oxide pollutants in the atmosphere, are likely contenders for inclusion in weathering equations. In Table 5.2(b), ΔG_f° calculations are given for a carbonic acid influence on the hydration of anorthite and albite. The energetics of these reactions are now quite different from those outlined for simple hydration in equations (3) and (4). Under the influence of carbonic acid, both anorthite and albite weather spontaneously at standard temperature and pressure. Since the (albite→ kaolinite + quartz) reaction yields a larger negative ΔG_r° value than the (anorthite→kaolinite) reaction, albite will weather preferentially to anorthite in soil systems, given the more realistic carbonic acid input.

It is interesting to speculate on the implications thermodynamic calculations have for the effect of acid rain on weathering rates of rocks and soil minerals. There is much evidence to suggest that the soil leaching rate of basic cations such as Ca^{2+}, Mg^{2+}, K^+, and Al^{3+} is very much increased under the influence of acid precipitation (see, for example, Reuss, 1980), although rarely have weathering reactions and rates been studied directly. In the River Elbe catchment, Paces (1985) calculated a doubling in the rate of chemical weathering of gneisses associated with an approximately ten-fold increase in sulphur dioxide in rainfall. These examples illustrate the potential for thermochemical modelling of environmental pollution effects on chemical weathering reactions and on soil processes generally.

Chemical tables can be consulted to obtain the ΔG_f° values at standard temperature and pressure (STP = 25 °C and 100 kPa) for a wide range of environmental compounds, including soil minerals (see, for example, Garrels and

Christ, 1965; Sadiq and Lindsay, 1979). Assuming that realistic weathering equations can be formulated, and also assuming that calculations at STP can be applicable to field soil conditions, the relative likelihoods of soil minerals weathering to produce known end products may be predicted. Clearly, changes in soil temperatures and pressures, due for example to unloading, geothermal energy or regional climatic differences, will influence the order and rate of chemical weathering reactions. Given that soil minerals weather differentially under the influence of uniform environmental conditions, the unequal rates of breakdown result in relative enrichment of stable minerals such as quartz and the iron oxides, and removal of less stable minerals such as olivine and calcite. An ideal, steady state soil profile condition is envisaged by Curtis (1975) when the input rate of fragmented parent material to the base of the profile equals the removal of soluble breakdown products from the whole profile. In the light of Figure 5.2 we can see that this is very much a geomorphological view of soil development, and that a more realistic view of soil formation and development processes requires three further considerations:

(a) the energetics of *in situ* chemical transformations of soil nutrients, other than processes producing new soil minerals;
(b) the energetics of cation and anion exchange processes on soil colloidal surfaces, and
(c) the energetics of soil organic matter decomposition.

In this short discussion, only decomposition processes will be considered further.

5.2.2   Energetics of organic matter decomposition in soil

In all sites with vegetation cover, a combination of mineral inputs from rock weathering processes and organic inputs from biological decomposition processes produces a recognizable soil material which is in a continual state of evolution.

Organic matter decomposition is a general term for a whole sequence of very complex processes whereby soil organisms use soil organic compounds as a food source. All living organisms, including those in the soil, are classified according to their main energy source into phototrophic (using solar radiation) or chemotrophic (using energy released from chemical reactions). Organisms can be further subdivided on the basis of their principal carbon source for nutrition into (a) autotrophs which use inorganic carbon sources (CO_2 gas), and (b) heterotrophs which use organic carbon compounds (such as carbohydrates, proteins, amino acids). Two *main* groups of living organisms result:

(a) Green plants which are *photoautotrophs*, gaining energy from the sun and carbon for nutrition from the intake of CO_2 during photosynthesis.

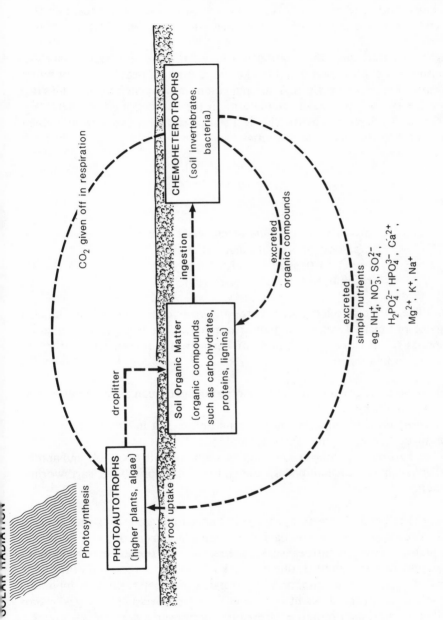

Figure 5.4 Energy relationship between plants (photoautotrophs) and soil organisms (chemoheterotrophs) during soil organic matter decomposition

(b) Animals, most soil organisms and man which are *chemoheterotrophs*, gaining both energy and carbon for nutrition from the breakdown or metabolism of organic carbon molecules such as carbohydrates.

Figure 5.4 illustrates the relationship between these two main groups of organisms in the soil system. Food intake and respiratory heat loss are the major pathways of energy flow through animal communities generally and the same is true for the soil organism population. In the soil community, the energy balance can be described by the classical simple equation used for both single organisms and for whole communities:

$$P = I - E - R \qquad (6)$$

where

$$\begin{aligned}
P &= \text{productivity (storage of chemical energy)} \\
I &= \text{ingestion (intake of chemical energy)} \\
E &= \text{excretion (loss of chemical energy)} \\
R &= \text{respiration (loss of heat energy)}
\end{aligned}$$

The difference between ingested energy and excreted energy is the assimilated energy or productivity of the system or organism. The basic energy equation (5) applies to all three main groups of heterotrophs represented in the soil organism population:

(a) *herbivores*, grazing on plants, roots and algae (such as some millipedes, slugs, woodlice);
(b) *carnivores*, preying on soil mites and other soil invertebrates (such as centipedes, some beetles, spiders); and
(c) *saprovores*, ingesting dead organic material of both vegetable and animal origin (such as some millipedes, some mites, some springtails, earthworms, fungi).

Like the soil mineral fraction, change in a specific organic substrate through time by decomposition is accompanied by an increase in entropy and a decrease in available energy for further work. The reason that the soil ecosystem does not 'run down' through time, due to a lack of food resources, is that the wide variety of organisms are supported by frequent and heterogeneous inputs of organic matter, primarily as litterfall and organism faeces. The largest group of soil organisms, incorporating the greatest species variety, are the saprovores, using a very wide range of dead organic materials as a food substrate. We shall use this group and the process of organic matter decomposition to examine the thermodynamics of the soil ecosystem.

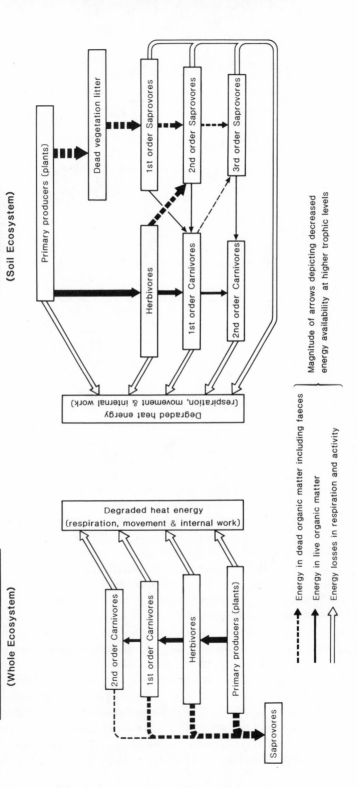

Figure 5.5 Simplified energy flow diagrams for: (a) ecosystems generally, and (b) the soil ecosystem

Table 5.3   Calorific values of plant, litter and soil materials

(a)	Soil organic constituents	Material	Calorific value (kJ g^{-1} ash-free dry weight) (references in brackets)
		Cellulose	17.6 (Jenkinson, 1981)
		Starch	17.5 (Jenkinson, 1981)
		Lignin	26.4 (Runge, 1973)
		Plant lipid	38.5 (Jenkinson, 1981)
		Resin, fat	36.8 (Runge, 1973)
(b)	Plant litters	Oak leaves	21.0 (Gorham and Sanger, 1967)
		Bryophytes	17–19 (Allen *et al.*, 1974)
		Pteridophytes	20–21 (Allen *et al.*, 1974)
		Gymnosperms	23–25 (Allen *et al.*, 1974)
		Grasses	19–21 (Allen *et al.*, 1974)
		Herbs (aerial parts)	20–21 (Allen *et al.*, 1974)

Energetics of ecosystems, including the soil ecosystem, can be illustrated in a trophic diagram in which energy is passed down the consumer chain (Figure 5.5). In all ecosystems, activity and respiration cause large losses of degraded thermal energy at each tier of the consumer community. Thus, less and less energy is available for further work at higher trophic levels. In 'normal' ecosystems, the largest energy stores and flows occur in primary producers and herbivores, while in the soil ecosystem, the largest energy stores and flows occur in the soil organic matter (dead primary producers) and the saprovores (Figure 5.5a).

The wide range of organic materials and molecules in the soil are degraded at different rates. Soluble compounds such as sugars and some amino acids and easily decomposable substrates such as proteins and carbohydrates are quickly used as a food source by a wide variety of organisms. Since more resistant organic compounds tend to have higher calorific values (Table 5.3), an ability to attack these materials would clearly be beneficial to decomposer organisms. Nevertheless, only a few specialist organisms can degrade lignin, lipids and chitin and the restricted numbers of specialist organisms leads to an accumulation of these resistant compounds, particularly in organic soil horizons. A comparison of plant litter and soil organic matter compositions (Figure 5.6) indicates that carbohydrates are easily decomposed (large amounts in plant litter, small amounts in soil) while lignin, resins and waxes are not decomposed or slowly decomposed (small amounts in plant litter, large amounts in soil). The lignin and fibre contents of different plant litters have been used as an index of decomposition rate. Van Cleve (1974), for example, found a negative correlation between lignin content of tundra litter types and rates of decomposition. The differing abilities of soil organisms to attack organic compounds leads to the view many soil biologists have of a succession of colonizing saprovores degrading

PLANT LITTER   SOIL ORGANIC MATTER

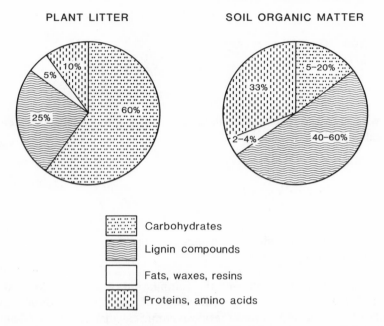

      Carbohydrates

      Lignin compounds

      Fats, waxes, resins

      Proteins, amino acids

Figure 5.6　Organic composition of plant litter and soil organic matter (Data averaged from Waksman (1938) and Kononova (1961))

an organic matter substrate, each wave of colonization altering the substrate for the next group of organisms, analogous to autogenic succession (Gray and Williams, 1971). Progressive depletion of the chemical energy resource occurs at each stage in colonization. Although the largest group of 'general' decomposer organisms have available to them a large energy resource of simple organic compounds, the very small group of specialist decomposers have available to them a potentially much larger energy resource of complex organic compounds.

If we wish to calculate the energetics of biological decomposition reactions thermodynamically, the procedure is slightly more complicated than for chemical weathering reactions. The standard technique used to measure the internal energy of complex organic materials is to measure their heat of combustion, or calorific content, using a bomb calorimeter. The enthalpy (ΔH) or potential energy can be approximated for most thermodynamic purposes from the heat of combustion. Since there is no direct way of measuring the Gibbs free energy (ΔG) for biological materials, the standard method is to calculate values from:

$$\Delta G = \Delta H - T\Delta S \tag{7}$$

where

$$\Delta G = \text{Gibbs free energy}$$
$$\Delta H = \text{enthalpy}$$
$$T = \text{temperature (}^\circ\text{K)}$$
$$\Delta S = \text{entropy}$$

A knowledge of the specific heat data for biological materials allows entropy to be calculated from:

$$\Delta S = \int_0^T \frac{C_p}{T}\, dt \tag{8}$$

where

$$C_p = \text{heat capacity at constant pressure}$$
$$t = \text{time}$$

These calculations have allowed the compilation of tables of ΔG_f° values at 25 °C and 1 atmosphere for a wide range of organic compounds. In the biochemistry of decomposition, thermodynamic calculations aim to determine the amount of energy liberated, or that obtained by the decomposers, during respiration. Chemoheterotrophs obtain energy either by aerobic respiration when the soil is well aerated, or by anaerobic fermentation when the soil is waterlogged. A comparison of the amounts of energy derived by these different processes during the degradation of glucose (Table 5.4) indicates that in chemically equivalent reactions, oxidation or aerobic respiration will always yield more energy than fermentation. This in turn means that many more microbial cells can be synthesized per unit of carbon substrate degraded (in this case glucose) when the soil is well aerated compared to waterlogged conditions. We can now understand why larger numbers of soil organisms, representing more species, inhabit well-aerated soils, resulting in fast rates of decomposition, while organic matter tends to accumulate, often forming peat, in waterlogged soils.

While calculations such as those in Table 5.4 allow the theoretical comparison of energy liberation for pure organic substrates at STP, equations cannot be written for the very complex reactions occurring when soil organic matter, plant litter or animal faeces decay. MacFadyen (1971) suggested a simple empirical technique for studying the energy derived by soil microbes decomposing complex organic substrates. The amount of CO_2 evolved during respiration from an enclosed soil or litter sample, or from an enclosed gas sampling tube, or chamber, in field soil is measured. Theoretically,

$$E_c = \frac{H}{C} \text{ for all organic compounds} \tag{9}$$

Table 5.4 Comparison of energy derived from (a) aerobic respiration, and (b) anaerobic fermentation of pure glucose. (All ΔG_f° values for organic compounds at 25 °C and 1 atmosphere, expressed in kJ mol^{-1}. ΔG_f° values taken from Metzler, 1977)

(a) Aerobic respiration of glucose

Oxidation equation:

$$C_6H_{12}O_{6(aqueous)} + 6O_{2(aqueous)} \rightarrow 6\ CO_{2(aqueous)} + 6H_2O_{(liquid)}$$
glucose

Free energy:

$$\Delta G_r^\circ = \Sigma \Delta G_f^\circ(6CO_2 + 6H_2O) - \Sigma \Delta G_f^\circ(glucose + 6O_2)$$
$$= [6(-394.4) + 6(-237.2)] - [(-917.2) + 6(0)]$$
$$= [-3789.6] - [-917.2]$$
$$\Delta G_r^\circ = -2872.4 \text{ kJ mol}^{-1}$$

(b) Anaerobic fermentation of glucose to ethanol

Fermentation equation:

$$C_6H_{12}O_{6(aqueous)} \rightarrow 2\ CO_{2(aqueous)} + 2CH_3CH_2OH_{(aqueous)}$$
glucose   ethanol

Free energy

$$\Delta G_r^\circ = \Sigma \Delta G_f^\circ(2CO_2 + 2\ ethanol) - \Sigma \Delta G_f^\circ(glucose)$$
$$= [2(-394.4) + 2(-181.5)] - [(-917.2)]$$
$$= [-1151.8] - [-917.2]$$
$$\Delta G_r^\circ = -234.6 \text{ kJ mol}^{-1}$$

where

$E_c =$ energy content per gram of carbon
$H =$ heat of combustion per gram of substance (kJ g^{-1})
$C =$ carbon content of substance (expressed as a decimal)

and since: 22.4 litres of CO_2 are chemically equivalent to 12 g carbon, then

$$E_{co_2} = \frac{0.536\ H}{C} \tag{10}$$

where

$E_{co_2} =$ energy liberated in the evolution of 1 l of CO_2

This simple calculation allows comparison of the amount of energy liberated during decomposition of any organic substrate, pure or complex (Table 5.5). *In situ* measurements of soil and litter respiration tend to yield excessively high CO_2 energy values. This is due to the inclusion of CO_2 evolved by root respiration. In woodland soils, root respiration alone can account for up to 50 per cent of the evolved soil CO_2. In decomposition energy studies it is now

Table 5.5   Energy liberation during decomposition of pine litter
and carbohydrate (from MacFadyen, 1971)

Pine litter	H	$= 20.02$ kJ g^{-1}
	C	$= 0.482$ (48.2%)
	E_{CO_2}	$= 22.26$ kJ l^{-1} evolved CO_2
Starch or cellulose	H	$= 15.65$ kJ g^{-1}
	C	$= 0.40$ (40%)
	E_{CO_2}	$= 20.97$ kJ l^{-1} evolved CO_2

H, C and E_{CO_2} are as defined in Equations (8) and (9)

more common to examine respiration during the decay of individual litter types as well as the total soil respiration. Studying the decomposition of tundra vegetation litter types, Flanagan and Veum (1974) found highest amounts of respired CO_2 from grass litters and lowest amounts of respired CO_2 from woody *Calluna* stems and shoots. These results confirm the earlier suggestion that fewer organisms can benefit from the energy yield of fibrous and lignified litter materials.

Since decomposer organisms require nutrients as well as energy for growth and reproduction, they take in, or *immobilize*, nutrients in their body tissues. Only nutrients surplus to their own requirements are excreted, or *mobilized*. Litter materials high in required nutrients, such as nitrogen (N), phosphorus (P) and potassium (K), are decomposed quickly. The decomposability of different litter types can be assessed on the basis of their C:N (total carbon: total nitrogen) ratio. Litters with low C:N ratios tend to be readily decomposed. The ratio of carbon to nitrogen, or to total amounts of other nutrients generally, can be described as the ratio of available energy to available nutrients for decomposer organisms. This concept is the crux of organic matter decomposition and nutrient cycling in soils.

5.3   ENERGETICS OF DYNAMIC SOIL PROCESSES

The movement of soil particles, soil water and solutes within soil profiles are controlled by four environmental gradients:

(a) the flow of all mass, including soil particles and water, occurs under the influence of *gravity*;

(b) the flow of water is also controlled by the development of a soil water head, or *soil moisture potential gradient*;

(c) the flow of salts in solution occurs under the influence of a *concentration gradient*; and

(d) the flow of heat occurs under the influence of a *thermal gradient*.

Individually, we can imagine any one of these transport mechanisms attaining equilibrium over small distances in the soil. However, both increasing soil distances and the interaction of any one gradient on other transport mechanisms, such as the effects of heating and cooling, or gravity, on the movement of soil water, mean that equilibrium is rarely, if ever, attained in field soils.

It is in soil hydrology that the most comprehensive treatment of the thermodynamics of dynamic soil processes has taken place. A very wide range of thermodynamic theory relating to the movement and changing state of water (liquid, ice and vapour) in field soils was presented by Edlefsen and Anderson (1943). Soil water behaviour is considered in terms of the changes occurring in its total free energy status. When temperature and pressure are held constant, the free energy of soil water is described by the total soil moisture potential (ϕ_t) which is the sum of the water potential (ϕ_w) and the gravity potential (ϕ_g) such that:

$$\phi_w = (\phi_m + \phi_o + \phi_p) \tag{11}$$

and

$$\phi_t = \phi_w + \phi_g \tag{12}$$

where

$\phi_m =$ matric potential; attraction of soil particles for water molecules means that energy must be expended to withdraw water.

$\phi_o =$ osmotic potential; attraction of salts in the soil solution for water molecules means that energy must be expended to withdraw water.

$\phi_p =$ pressure potential; when the pressure acting on soil water is different from the reference state (or standard pressure of 100 kPa), work will have to be done in transferring a quantity of water from the standard state to the soil pressure. Work is positive if soil pressure is greater than standard pressure and negative if it is less. The pressure below the water table in waterlogged soil is always greater than 100 kPa.

When the soil is unsaturated, work must be done to extract water from the soil and all three potentials making up ϕ_w have negative ΔG values. Only the gravity potential has a positive ΔG value which increases with height above a given reference level. Since the overall ΔG for soil water, obtained from: $\phi_t = \phi_w + \phi_g$, is negative under these circumstances, water is retained in the soil. Only when the soil is saturated does ϕ_t become positive. $\phi_w + \phi_g$ can then result in a positive value for ΔG and water drains out of the soil.

Clearly the thermodynamics of soil moisture behaviour in field soils are complicated by the operation of thermal and pressure gradients. It is also important to appreciate that the two examples described above represent extremes of soil moisture condition. The movement of soil water when the soil is unsaturated is controlled mainly by the relative magnitudes of the matric and pressure potentials. The magnitude of ϕ_m increases with increasing clay and organic matter content, so that any increase in clay or organic matter increases the amount of water that can be retained against the force of gravity. The development of soil horizons roughly parallel to the soil surface results from the downward translocation of clay and organic particles under fluctuating drainage conditions.

The location of a soil on a slope or of a particular horizon within a soil profile thus imparts potential energy, due to its position in the gravity force field. This energy is translated into kinetic energy of movement, usually with water acting as the transporting agent to distribute soil components downslope and downprofile. Two examples of this kinetic energy are:

(a) erosion, resulting in removal of soil A horizon material from upslope with deposition of new A horizon material downslope; and
(b) the formation of distinct horizons such as ironpans and clay pans due to the downprofile translocation of organically complexed iron compounds and clay colloids respectively.

5.4   ENERGY CHANGES DURING SOIL DEVELOPMENT

As soil development proceeds, changes occur in the relative contribution of the three different forms of energy in the soil (see Table 5.1). If we assume that both organic and inorganic soil resources are finite (that is, there are no inputs of weatherable parent materials or decomposable vegetation litter), the contribution of total chemical energy to the overall energy status of the soil will increase to a peak when maximum weathering and decomposition occur but will then decline as these resources are used up. Similarly, the importance of mechanical energy in soil development declines with time as land denudation occurs and as all weathered and translocatable particles are moved downprofile. Three conditions are vital for maintaining high mechanical and chemical energy status in the soil:

(a) rejuvenation or uplift;
(b) any geomorphological process providing new weatherable mineral material (examples include: innundation and flooding resulting in sedimentation; vulcanism, producing lava and ash; or erosion and deposition); and
(c) the development of a large, dynamic vegetation cover providing litter fall, root death and encouraging a large, active soil organism population.

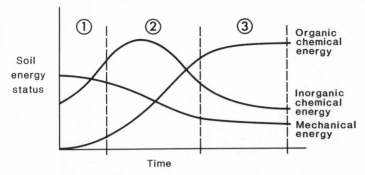

Figure 5.7 Three energy stages in the development of soil from the exposure of parent material to the development of 'climax' (equilibrium vegetation) (see text for explanation)

The relative contributions of chemical and mechanical energy to the more realistic soil developmental sequence from the exposure of parent material to the development of a so-called 'climax' or equilibrium vegetation is illustrated in Figure 5.7. Three apparent energy stages can be identified:

Stage 1 Initially, with the exposure of bare rock, mechanical energy is the main input to soil development, as physical weathering exposes a larger rock surface area for the processes of chemical weathering. This results in an increase in the importance of inorganic chemical energy as more mineral crystals and molecules take part in chemical weathering reactions. The colonization of vegetation and initiation of a vegetation succession add litter to the soil and increase soil organism numbers, resulting in a rise in importance of organic chemical energy.

Stage 2 A peak in inorganic chemical weathering occurs as all easily weathered minerals are exhausted. Increasing vegetation colonization and microbial activity result in a steady rise in importance of organic chemical energy. The importance of mechanical energy declines as rock disintegration proceeds, particle translocation downprofile proceeds and plant roots improve soil stability.

Stage 3 When an equilibrium vegetation assemblage develops, a regular litter input maintains the organic chemical energy status at a high level. The actual level is determined by the balance between the rate of litter input and the rate of organic matter decomposition. The importance of inorganic chemical energy declines to a low, steady level, dependent on the balance between the availability of new, weatherable parent material and rates of weathering. Mechanical energy is maintained at a steady, low level, represented in maturing soil profiles by the downward and lateral movement of water, particles and solutes in processes such as eluviation/illuviation, leaching and soil erosion.

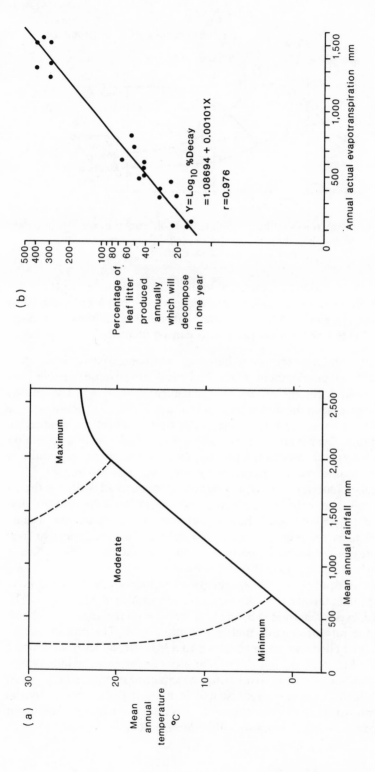

Figure 5.8 Relationship between climatic parameters (temperature, precipitation, evapotranspiration) and (a) chemical weathering (redrawn with permission from Wilson, 1968), and (b) organic matter decomposition. Correlation and regression of actual annual evapotranspiration with annual weight loss of leaf litters (redrawn with permission from Meentemeyer, 1978). At a decomposition rate of 100 per cent, all of the available leaf litter is decomposed in 1 year. At a decomposition rate of 200 per cent, all leaf litter is decomposed in 6 months

Rates of mechanical and chemical processes control the duration of these three energy stages of soil development. Within the soil, two major limiting factors control these rates: (a) supply of mineral and organic resources (reactants); and (b) transport of inorganic and organic reactants to and products from the reaction site. Assuming a steady supply and transport of reactants, soil temperatures, controlled by solar energy receipt and transmission, are the main influence on process rates. Higher temperatures speed up reaction rates by increasing both the energy and the probability of particle collisions. Over the temperature ranges commonly found in soils (0–30 °C), both inorganic, chemical reactions and microbially mediated, biochemical reactions generally increase by about two to three-fold for every 10 °C rise in temperature (Wiant, 1967; Bohn, McNeal and O'Connor, 1979). This relationship between reaction rate and temperature is called the Q_{10} value and is given by:

$$Q_{10} \frac{\text{rate at } t + 10 \,°C}{\text{rate at } t \,°C} \tag{13}$$

where

$$t = \text{ initial temperature } (°C)$$

On a world scale, geographical differences in solar radiation receipt and the ratio of rainfall to evaporation have been widely used to compare rates of chemical weathering and organic matter decomposition. The examples illustrated in Figure 5.8(a) and (b) suggest that both processes are at a maximum in hot and wet conditions, such as in tropical soils, declining to a minimum in cold and dry conditions such as those found in semi-arid continental interiors. Meentemeyer (1978) found a very good correlation between decomposition rate, as measured by percentage weight loss of leaf litters and actual evapotranspiration which embodies both solar energy and moisture availability (Figure 5.8b)).

5.5 CONCLUSIONS

The energetics of environmental systems have interested man for three main reasons: (a) to harness energy for his own use; (b) to predict environmental hazards, or occasions on which energy in the environment may become damaging to him and to his infrastructure (examples include landslides, hurricanes, floods); and (c) to determine efficiency and productivity of biological systems potentially useful to him in the production of food. Although the energetics of soil systems have rarely, if ever, attracted attention on any of these counts, they clearly play a very important role in agriculture (see (c) above).

Agricultural systems can be considered as particularly simple ecosystems. The development of highly productive agricultural systems relies on large nutrient inputs to the soil in the form of inorganic fertilizers or organic manures. We can think of these additions as (a) inputs of inorganic chemical energy, analogous to new weatherable rock minerals, and (b) inputs of organic chemical energy, analogous to new decomposable plant litter. Agronomists speed up the inputs to the 'boxes' in Figure 5.2, but also speed up to the outputs. In agricultural soils, both inorganic and organic 'boxes' act as open systems, with large losses of simple nutrients, and hence chemical energy, occurring through leaching to drainage waters and in both plant and animal crops when they are harvested.

While the scale of energy transformations in soil systems is very much smaller than that of atmospheric, geomorphological or ecological systems, the modes of operation of these energy changes are entirely analogous to other branches of environmental science. The thermodynamics of chemical weathering, organic matter decomposition and soil transport processes are similar in operation to those of chemical, biological and physical systems respectively.

References

Allen, S. E., Grimshaw, H. M., Parkinson, J. A., and Quarmby, C. (eds) (1974). *Chemical Analysis of Ecological Materials*, Oxford: Blackwell Scientific Publishers.
Bohn, H. L., McNeal, B. L., and O'Connor, G. A. (1979). *Soil Chemistry*, Chichester: John Wiley and Sons, 329 pp.
Curtis, C. D. (1975). Chemistry of rock weathering: fundamental reactions and controls. In Derbyshire, E. (ed.), *Geomorphology and Climate*, Chichester: John Wiley and Sons, 25–57.
Curtis, C. D. (1976). Stability of minerals in surface weathering reactions: a general thermochemical approach. *Earth Surface Processes and Landforms*, **1**, 63–70.
Drever, J. I. (1982). *The Geochemistry of Natural Waters*. Prentice-Hall, 388 pp.
Edelfsen, N. E. and Anderson, B. C. (1943). Thermodynamics of soil moisture. *Hilgardia*, **15(2)**, 31–298.
Flanagan, P. W. and Veum, A. K. (1974). Relationship between respiration, weight loss, temperature and moisture in organic residues on tundra. In Holding, A. J. *et al.* (eds). *Soil Organisms and Decomposition in Tundra*. International Biological Programme, Tundra Biome, 249–277.
Garrels, R. M. and Christ, C. L. (1965). *Solutions, Minerals and Equilibria*. London: Harper and Row.
Gorham, E. and Sanger, J. (1967). Calorific values of organic matter in woodland, swamp and lake soils. *Ecology*, **48**, 492–494.
Gray, T. R. G. and Williams, S. T. (1971). *Soil Micro-organisms*. Edinburgh: Oliver and Boyd.
Jenkinson, D. S. (1981). The fate of plant and animal residues in soil. In Greenland, D. J. and Hayes, M. H. B. (eds). *The Chemistry of Soil Processes,* Chichester: John Wiley and Sons, 505–561.
Kononova, M. M. (1961). *Soil Organic Matter*. Oxford: Pergamon Press.
MacFadyen, A. (1971). The soil and its total metabolism. In Phillipson, J. (ed.) *Methods of Study in Quantitative Soil Ecology*. International Biological Programme, Handbook No. 18, 1–13.

Meentemeyer, V. (1978). An approach to the biometeorology of decomposer organisms. *International Journal of Biometeorology*, **22(2)**, 94–102.

Metzler, D. F. (1977). *Biochemistry. The Chemical Reactions of Living Cells*. London: Academic Press.

Paces, T. (1985). Sources of acidification in Central Europe estimated from elemental budgets in small basins. *Nature*, **315**, 31–36.

Reuss, J. O. (1980). Simulation of soil nutrient losses resulting from rainfall acidity. *Ecological Modelling*, **11**, 15–38.

Runge, M. (1973). Energieumsatze in dem Biozonosen Terrestrischer Okosysteme. *Scripta Geobotanica*, **4**, 77–89.

Sadiq, M. and Lindsay, W. L. (1979). Selected standard free energies of formation for use in soil science. In Lindsay, W. L. *Chemical Equilibria in Soils*. Chichester: John Wiley and Sons, 386–413.

Van Cleve, K. (1974). Organic matter quality in relation to decomposition. In Holding, A. J. *et al.* (eds). *Soil Organisms and Decomposition in Tundra*. International Biological Programme, Tundra Biome, 311–324.

Waksman, S. A. (1938). *Humus: Origin, Chemical Composition and Importance in Nature*. Baltimore: Williams and Wilkins.

Wiant, V. W. (1967). Influence of temperature on the rate of soil respiration. *Journal of Forestry*, **65**, 489–490.

Wilson, L. (1968). Morphogenetic classification. In Fairbridge, R. W. (ed.) *Encyclopedia of Geomorphology*, Volume 3. Rheinhold, 717–729.

Energetics of Physical Environment
Edited by K. J. Gregory
©1987 John Wiley & Sons Ltd

6

Order and Truth:
Energetics in Biogeography

I. G. SIMMONS

Department of Geography, University of Durham

6.1 THE NATURE OF BIOGEOGRAPHY

The term 'biogeography' is not uniquely used by geographers. It is in use by biologists as well and the two groups have attached various meanings to the term. It seems useful, therefore, to begin with a brief exploration of the word itself. As used by biologists, it has come to mean the study of the distribution of plants and animals at a variety of scales (usually excluding the individual stand of vegetation for instance but thereafter extending to global patterns of presence and absence) and in a temporal framework which deals with at least hundreds and perhaps many thousands of years, that is, evolutionary time. Some geographers who have engaged in the study of plants and animals have adopted this approach (see, for example, Seddon, 1971; Stott, 1981). In these studies, however, other geographers have become more interested in approaches to the living world which are characterized by the science of ecology. This viewpoint emphasizes the interactions of living organisms between each other and their non-living environment, and takes place on a relatively limited time-scale of not more than hundreds of years, that is ecological time. Practitioners may engage, for example, in studies of the distribution and internal patterning of vegetation types, or of the way in which plant and animal succession follows disturbance by natural events or by human intervention, or in problems of the management of natural or nearly natural environments for conservation purposes (Watts, 1971; Kellman, 1980; Tivy, 1982). In many cases the work of geographers is little different from that of biologists; the example of Quaternary ecology may be quoted as (a) one bringing together both distributional and ecological approaches, and (b) one in which geographers' work is no different from (and often carried out in conjunction with) other earth scientists.

One of the approaches of biologists most adopted by geographers in recent years has been a functional approach to ecology. This is concerned with the relationships between the various components of a biological community and derives in particular from the idea that this community is a dynamic entity exchanging energy and matter with its surroundings. The concept of the constituents together with these interactions, makes them conformable to the general definition of a system, and the term 'ecosystem' formally defined by E. P. Odum (1959) as 'any area of nature that includes living organisms and non-living substances interacting to produce an exchange of materials between the living and non-living parts is an ecological system or ecosystem', has been widely used in the English-speaking world; biogeocoenosis has something of the same meaning in Russian language studies. In trying to understand the behaviour of ecosystems in a quantitative (and eventually predictive) way, ecologists have measured such characteristics of ecosystems as the flow and cycling of nutrients like nitrogen, calcium or phosphorus, the dynamics of the populations of the various organisms which make up the system, and most emphatically the flow of energy through the ecosystem, a field of enquiry known as ecological energetics. Since the same measurements can often be made when human activity is an important agent of change in the ecosystem, geographers have seen this functional approach (including but not confined to energy flow) as a potential linkage between human and physical geography and have incorporated it into their teaching (Simmons, 1979, 1981). Not many, it must be said, have carried out primary research of this kind. Thus the study of energy flow through ecosystems is one in which the results have been to some extent used in geography but the primary data have been gathered outside that discipline (Stoddart, 1965, 1986).

6.2   ENERGETICS IN ECOLOGY

The genesis of the realization of the dependence of organic life upon the energy provided by the sun is lost in antiquity. The gradual refinement of the connections between sunlight, water, the minerals of the earth and living things can be reconstructed in writings as diverse as the Norse creation myths, Greek botanical philosophy, the homilies of St Basil, Nicolas of Cusa, and the German chemist J. von Liebig. When we come to this century (Lieth, 1978), the ideas begin to crystallize clearly around the notion of feeding relationships. Here is clearly a critical (even if not all-important) dimension of the structure of nature and one in which energy is an indispensable element. Indeed, organisms could be seen simply as devices for processing solar energy to the point where it could be radiated back to space as heat. So the concept of a food chain underlies much of the early scientific study of ecological energetics: from the sun, energy may pass to a plant, which is eaten by an animal; this organism is in turn eaten by another animal and there might be yet another animal at the end of the chain.

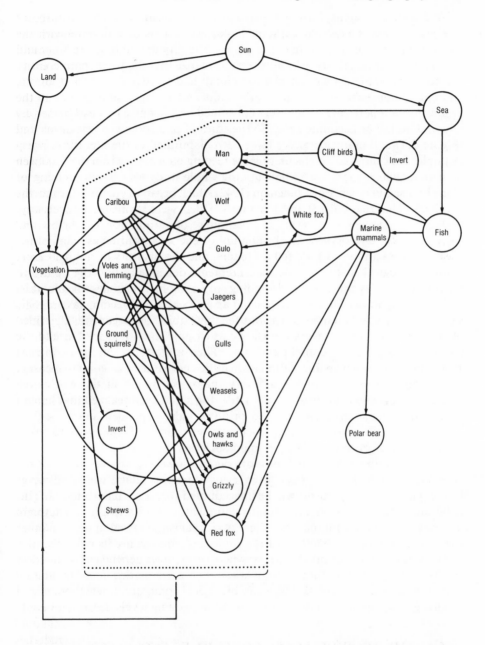

Figure 6.1 A food web for the Arctic tundra; on land the herbivores are dominated by three groups which are the food supply of a number of carnivorous species (Source: Simmons, 1979)

In terrestrial systems, however, three or four steps are normal, for instance sun–grass–mice–snakes–hawks; some biologists have thought that the seas may contain larger chains since the first plants (the phytoplankton) are so small. It took little serious fieldwork, however, to realize that the food chain concept was too simple. First, many species of animals had more than one food source, and indeed some ate for example both plants and animals; second, not all the organisms at a particular stage were consumed and in any case dead biological material had to be accounted for. So the concept of a food web was developed (Figure 6.1) and many examples have been elaborated in the literature. Early examples concentrated on elucidating the feeding pathways within the food web by naming which species ate what, but the whole study was placed on a higher plane by a seminal paper of Lindeman (1942) which began to quantify the process involved. This sparked off investigations which were to measure the energy budget of ecosystems in terms of incoming solar radiation of the correct wavelength for photosynthesis, the amount of energy appearing as plant tissue, the amount as animals which fed on plants, and the quantity of animal-eating animal biological material. The measurements depended upon the ability to measure the energy content of the standing crop of biological tissues (the biomass) and the segregation of the components of an ecosystem into feeding layers in the sun, the plants, and animals of various feeding habits, the organisms that lived off dead tissues. These layers are known as trophic levels and are basic to the measurement of energy content and flow. Their importance culminated in the 1960s and 1970s when an international scientific endeavour was set up to measure the energy flows and contents of every major ecosystem in the world: the International Biological Programme (IBP), for it had been realized that human welfare in its most basic aspect of nutrition was also dependent upon the flow of energy through ecosystems, even though some of the systems were manipulated or even created by human societies rather than being natural.

Before describing such work in more detail, however, we need to enter the caveat that a model of an ecosystem in terms of trophic relationships will never be complete: some organisms will be unaccounted for, some links omitted. This is because there are other important relationships besides feeding. These are interrelationships like mutualism, where two organisms depend upon each other for their existence (the bacteria of the ruminant stomach are its chief features but they could not exist outside it), or where one species depends upon another for its dispersal and in animal pollination. These are inadequately treated by trophic analysis, but nevertheless every biological community, whether natural or changed by humans, has within it the basic feeding levels described below.

6.3 THE TROPHIC LEVEL CONCEPT IN ECOSYSTEM STUDY

At its simplest, we can conceive of every ecosystem receiving an input of solar energy. Some of this is captured by plants and transformed to chemical energy which is manifested as plant tissue; this is the producer trophic level (*TL*). Animals eat some of this plant production: they are the herbivore *TL*, and in turn some are eaten by other animals which form the carnivore *TL*. If there are further levels they consist of top carnivores. Uneaten plant and animal material and corpses are processed by a rich mix of decomposer organisms; fungi and bacteria are common elements but the level is not confined to them.

If we elaborate this simple scheme a little further then we must start with the transformation of the radiant energy of the sun into the energy-rich chemically based organic matter of plants: the process of photosynthesis. This has an involved biochemistry but, simplified, occurs in two steps: the first uses light energy absorbed by chlorophyll to split a water molecule, releasing oxygen; the second, which does not require light, uses the energy in several steps to reduce carbon dioxide to carbohydrates. This can be summarized in the equation:

$$6CO_2 + 12H_2O \xrightarrow[\text{(chlorophyll)}]{\text{2816 kJ}} C_6H_{12}O_6 + 6O_2 + 6H_2O$$

Obviously, the availability of energy as light is fundamental to photosynthesis; as light intensity varies according to many factors such as time, season, weather, topography, orientation of the photosynthetic surface (usually a leaf in higher plants), shading by other biotic or abiotic structures and even the depth of water where aquatic plants are concerned. In general, the rate of photosynthesis increases linearly in relation to the intensity of light, but after a certain threshold value photosynthesis uses less and less of the light (that is, becomes less efficient) until above an intensity of 10 000 lux, an increase in light intensity produces no increase in photosynthetic rate. This plateau is called light saturation, and if the rate of photosynthesis is to be further increased then other factors have to be adjusted. In fact, except for plants growing in dense shade (for example, below closed forest canopy) there is normally enough light from the sun to saturate the photosynthetic capacity of most plants.

The end result of photosynthesis is a carbohydrate $C_6H_{12}O_6$ which can be converted by the plant into starch and stored; it can be combined with other sugar molecules to make cellulose which is a basic structural material in plants; or it can be combined with elements such as nitrogen, phosphorus and sulphur to produce proteins, nucleic acids and all the other constituents of living cells. Some of the sugar produced by photosynthesis is used as an energy source by the plants themselves for growth, the maintenance of tissues, and biochemical processes. This process is called respiration and can be summarized as:

$$C_6H_{12}O_6 + 6O_2 \longrightarrow 6CO_2 + 6H_2O + 2830 \text{ kJ}$$

It is worth noting that this energy is converted to heat in the course of its use by the plant, and so is never available for use again within the ecosystem since it is a dispersed low-grade energy rather than the concentrated high-grade chemical energy which is incorporated into plant tissues. The amount of energy available for the heterotrophs in the ecosystem (that is, the animals and the decomposer organisms) is therefore dependent upon the balance between the rates of photosynthesis and respiration.

In their energy relations, all food chains and webs are subject to one of the basic laws of physics governing energy flow, namely the second law of thermodynamics. Put simply, the law requires that at every transfer step in the ecosystem some energy will be degraded from a highly concentrated chemical form to a highly dispersed form as heat which cannot be recycled into chemical energy but must be radiated out of and lost to the ecosystem. So at each trophic level a conversion to heat takes place which means that less energy becomes biomass at the succeeding trophic level, especially the herbivore and carnivore levels. Consider as a simple example the foraging of a herbivore which means that a portion of its energy intake is expended in finding more food: a condition which is exacerbated if the herbivore is warm-blooded and has to expend a great deal of energy catching its prey and many are thus adapted to eating at relatively infrequent intervals, such as lions and snakes. One result of this law is that in a given ecosystem, the net availability of energy gets less at levels successively away from the producers: if the energy content (usually given as the calorific value) of each *TL* is plotted (Figure 6.2) then a pyramidal shape is obtained; very often the numbers of organisms also follow the same trends, although if the producers are very large (for example, oak trees) then they may well be exceeded in number by herbivores such as defoliating caterpillars. However, a simple grass field-mice–snakes–hawks ecosystem clearly follows the normal trends in this respect; the absolute number of predators is quite small compared with the number of grass plants. Contrariwise, the sizes of individual animals may increase up the chain since it may well be an advantage for a predator to be larger than its prey. Exceptions are not hard to find to this generalization, however. The ecosystems may appear to have more energy at their consumer

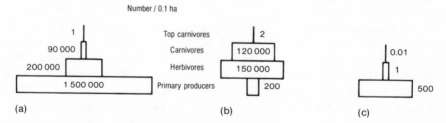

Figure 6.2 Number and biomass pyramids for ecosystems. (a) and (b) are pyramids of numbers; (a) of a grassland in summer, and (b) a temperate forest in summer where the producer level consists of trees that are large in size but few in number; (c) is a biomass pyramid (dry weight/unit area) for an abandoned agricultural field undergoing succession
(Source: Simmons, 1979)

stages than in their production; this can be so because organic matter is imported across the ecosystem boundary—for example, carried by running water into a pond, or by the tides into an estuary or by lunchtime office workers into a duck-inhabited park lake. These facts have obvious relevance for the human use of biota. If an ecosystem (whether natural or man-influenced) is being cropped for food, then there will be much more energy available per unit if man eats as a herbivore rather than as a carnivore, assuming that the plants are as edible as the animals.

Another way in which the energy and matter present in living organisms is made manifest is in the rate of production of organic matter. The fundamental process is the production of organic tissues incorporating solar energy by green plants, and its most important practical expression is net primary productivity (NPP) which is the material actually available for harvest by animals and for decomposition by the soil fauna and flora or their aquatic equivalents. NPP therefore integrates abiotic phenomena such as the non-uniform distribution of incident radiant energy and the different conditions of moisture supply, with biotic features such as the genetic properties of the plants which are the primary producers. It is usually measured as dry organic matter synthesized per unit area per unit time and expressed either as $g m^{-2} year^{-1}$ or $kg ha^{-1} year^{-1}$. It can also be expressed as the calorific value of the dry matter, in kilocalories (kcal) or joules ($1 cal = 4.2 J$); or in grams of carbon in the dry matter; this figure can reasonably be multiplied by 2.2 to obtain a dry-matter equivalent. At any one time, the standing crop of living organisms present per unit area is the biomass.

The ranking of the average NPP of the various biome types yields few surprises. (Table 6.1) The list is headed by the tropical rainforests which have a year-round growing season and a high biomass, so that they would perhaps be expected to produce the most organic matter in the course of a year. Not generally appreciated perhaps is that limited parts of the oceans such as estuaries and coral reefs reach the NPP of tropical forests. In these data they are outweighed in absolute terms by the immense areas of open ocean whose NPP is more like that of the tundra. Tropical grasslands (including grass-dominated savanna) overlap with some of the woodlands of less favourable climates, and there is a big gap between these ecosystems and those of tundra, desert scrub and desert. The position of agricultural land is of some interest. At a world average of $650 g m^{-2} year^{-1}$, it exceeds the average figure for grasslands but falls well below most of the forests. In one sense this is a misleading figure because in Western-style agriculture it is brought about by large inputs of energy derived from fossil fuels as well as solar energy. The column for total production emphasizes the role of the forests in providing the bulk of the NPP of the world (62.8 per cent of the continental area, 42.8 per cent of the total), whereas cultivated lands produce only about 7.7 per cent.

The efficiencies of energy transfer from plant to herbivore are often low. Thus in the tundra, the biomass of plants exceeds that of animals by $\times 15$, in deciduous forests × 300, and in coniferous forests × 1200. These data are to be expected given the metabolic needs of animals with regard to temperature and the operation of the thermodynamic laws within ecosystems. More to be emphasized at this point is that secondary production has two branches: (a) the herbivore–carnivore–top carnivore chain, where secondary production represents successive reorganization of the same molecules; and (b) the decomposer chain which begins with dead organic matter, where the sum of new organisms in the decomposer chain (such as fungi–bacteria–protozoa) is the secondary production. The efficiency of ingestion of food may be quite high: vertebrate herbivores may consume up to 25 per cent of available NPP of grasslands, and 5 per cent in forests. Invertebrate herbivores usually consume less than 5 per cent of above-ground primary production, although during population 'explosions' of, for example, locusts or caterpillars all the annual leaf production may be consumed; this still represents less than 50 per cent of NPP in grasslands and 20 per cent in forests although the overall system effect may be somewhat catastrophic. Predation by vertebrates upon invertebrates usually accounts for 26 per cent of prey mortality; the invertebrate prey of titmice and shrews in a wood suffered only 5 per cent of its mortality from these its predators. By contrast nearly all the mortality of wildebeest in East Africa resulted from predation by the spotted hyaena; 100 per cent of the herbivore production was in this case harvested by predators.

Although the large secondary producers of the herbivore–carnivore chain are the most obvious and the most discussed, the decomposer system must be given its due importance. In a typical grassland 86.5 per cent of the NPP energy passed to the saprovore system whereas only 13.5 per cent went to the herbivores. No wonder therefore that the productivity of the micro-organisms exceeds that of the animals; in many ecosystems there is more life below the surface than above it.

If it is now estimated with reasonable accuracy that total world NPP is about 170×10^9 tonnes year^{-1}, of which 50–60 tonnes year^{-1} is from the oceans, does this have a significance for our use of biota as a resource, bearing in mind for example the low proportion of world NPP contributed by agriculture (7.7 per cent of terrestrial NPP), and the low productivity of much of the oceans? First of all, it puts in perspective the activities of man. The energetic magnitude of world primary production is estimated at 6.9×10^{17} kcal year^{-1}, whereas man's use of fossil fuels and other industrial energy in 1970 was 4.7×10^{16} kcal year^{-1}, around 7 per cent of NPP. The concern is that the latter figure has been doubling every 10 years whereas the former may diminish from human impacts and that the impact of the industrial energy is not uniform; we need to know which high natural productivities are being affected by it, as with estuaries. Man's harvest of food also looks small compared with the biosphere production: about 0.72 per cent of the energy of global NPP. But the great abundance of wood, grass tissue and phytoplankton cannot be used without incurring heavy energy expenditure in the form of industrial fuels or wasteful animals.

Table 6.1 Net primary productivity (*NPP*) for the world values around 1950. (Source: Whittaker and Likens, 1975)

Ecosystem type	Area $(10^6\,km^2)$	Mean NPP $(gm^{-2}\,year^{-1})$	Total production $(10^9\,tonnes\,year^{-1})$
Tropical rainforest	17.0	2200	37.4
Tropical seasonal forest	7.5	1600	12.0
Temperate forest: evergreen	5.0	1300	6.5
Temperate forest: deciduous	7.0	1200	8.4
Boreal forest	12.0	800	9.6
Woodland and shrubland	8.5	700	6.0
Savanna	15.0	900	13.5
Temperate grassland	9.0	600	5.4
Tundra and alpine	8.0	140	1.1
Desert and semidesert scrub	18.0	90	1.6
Extreme desert: rock, sand, ice	24.0	3	0.07
Cultivated land	14.0	650	9.1
Swamp and marsh	2.0	3000	6.0
Lake and stream	2.0	400	0.8
Total continental	149	782	117.5
Open ocean	332.0	125	41.5
Upwelling zones	0.4	500	0.2
Continental shelf	26.6	360	9.6
Algae beds and reefs	0.6	2500	1.6
Estuaries (excluding marsh)	1.4	1500	2.1
Total marine	361	155	55.0
World total	510	336	172.5

What is disturbing is the effect of human societies in reducing all forms of productivity, by replacing highly productive systems like rainforests with grassland and scrub and by toxification of ecosystems with agricultural, industrial and urban wastes. Together these reduce the production of organic matter which is man's most important resource, since it is infinitely renewable.

The inclusion of agriculture in Table 6.1 shows us that the trophic level concept is useful in some studies of the human use and manipulation of the natural environment. In a broad sense, any study of the environmental impact of human activity can focus upon its effects at various trophic levels: big game hunting is likely to remove predators for example; poisoning offshore waters is likely to reduce the producer level by inhibiting photosynthesis by phytoplankton. Yet nowhere is the idea more relevant than that of the feeding levels of our own species. In poor societies (with notable exceptions like hunter-gatherers and pastoralists) very little meat is eaten, whereas in rich societies it becomes available and indeed symbolic—in the United States, for instance, 'real men don't eat quiche'. Simply, a poor society (especially if densely populated) cannot afford to divert energy into a higher, animal, trophic level. If meat is eaten then the pig may be very popular because it can function as part of the decomposer level since it will scavenge around for almost anything organic. One reason why rich societies eat more meat is that they have available to them cheap and reliable stores of non-solar energy in the form of fossil fuels which can subsidize the solar flows; this is elaborated below.

Most engineering treatises on energy stress the efficiency with which it is used — that is, the ratio between energy put into a device and useful energy output. In machines, the efficiencies calculated are of the order of, for example, 34 per cent for electricity generation at a thermal power station. In nature, we have already noticed that about 0.2 per cent of photosynthetically available radiation is utilized by plants. The transfer efficiency to further stages of food webs seems to be highly variable, perhaps between 0.1 and 25 per cent, with a concentration in the 7–14 per cent range, a latitude which may easily result from inaccuracy in measurement. Setting aside the fine detail of this argument, it looks as if efficiency is not a parameter for which evolution has tended to provide the maximum value — that is, evolution has resulted in the maximization of some other characteristic (reproductive efficiency or transmission of the 'best' genes, for instance) than the efficiency of energy transfer. Efficiency may therefore be a human characteristic rather than a natural one.

6.4   PROBLEMS WITH THE TROPHIC LEVEL CONCEPT

One fundamental objection raised by biologists to these ideas is simply that trophic levels are indeed concepts; they are not physical realities themselves and therefore cannot have properties. Thus hypotheses about a trophic level (as distinct from organisms themselves) could not be tested. There is also the publicity of trying to assign an organism to a trophic level. By definition each species is put one level above its food, but this requires us to know the trophic level of the food. Since many carnivores, for example, are opportunistic in their feeding, this may be virtually impossible; a hawk may recently have fed on small birds, some of which are seed-eaters, and also on other species which eat insects. Some of these insects may themselves be carnivorous, others herbivorous. The same hawk, therefore, may appear at the top of a food chain such as living plant–insect–small bird–hawk, or a chain like plant detritus–soil microfauna–small worms–insects–spiders–small birds–hawks. Another difficulty with the concept of trophic levels is the homogeneity attributed to organisms as food sources. Yet a plant may offer different tissues each with different palatabilities and energy contents: there are at least six different states of these, from root mucilage and leaf drip through to storage organs such as seeds. One more objection about trophic levels is that they are historically determined. They describe the history of energy flow but not the present state of the energy (which in being transformed does not keep its past history), nor what may happen tomorrow. As a response to these criticisms, ecologists are preparing a return to classifying ecosystems as pyramids of numbers (as first proposed by C. Elton in the 1920s) since size of animal is clearly paramount in feeding relationships. And we can expect the neatness of the trophic level concept to be blurred by more complex models (Cousens, 1980, 1985; Rigler, 1975), and our view of energetics in biogeography will itself need modification.

These particular objections relate to reconsiderations in a wider framework. The first of these is that the reduction of ecosystem complexity to a set of measurements of energy content appears to do away with the need for systematics and classification of organisms; we have, for example, plants *en masse*, or herbivorous animals (insects and red deer need not be separated out). Yet ecosystems are clearly not like that: they comprise real individual organisms with complex patterns of evolution and behaviour, and to reduce them to a model in which only kilojoules per square metre per year ($kJ\,m^{-2}\,year^{-1}$) are relevant is clearly to lose a lot. The American ecologist F. E. Egler, not one to miss the chance of an acerbic phrase, once remarked that the approach was a bit like grinding up cows to make hamburgers; you could not be certain that a monkey hadn't slipped in somewhere. Then again, while we might admit to the idea of trophic levels being a useful descriptive device (especially if the ecosystem is so simple that each organism can be confidently assigned to a single station in the structure of levels), it is less likely that the model possesses explanatory power, unless of course energy is always the limiting factor in ecosystem development and maintenance. Yet we know that this is not always so: maximization of efficiency seems not to have been selected for in the course of evolution; nutrients are often seen to be limiting and in any case many organisms limit their populations to a number below the absolute carrying capacity of their habitat. So while energetics in ecology may describe (though there are reservations about this), explanation is even more fraught with hazard.

6.5 LINKS WITH HUMAN GEOGRAPHY

The number of ecosystems on Earth which is unaffected by human activity (whether that be deliberate or accidental) is much diminished from what it was in say 9000 BC, or 1800 AD, and is daily getting fewer. So any biogeography which fails to take account of the human presence and its manipulative powers is something of a retreat to an Arcadian fantasy, since for millennia men and women have altered the genetic make-up of plants and animals, created new ecosystems and obliterated old ones. If, then, energetics has any role to play in the description and analysis of natural ecosystems (and we have seen that it has some value though hedged around with limitations) then it ought to be some value in similar studies of human-influenced systems.

In general we can perhaps say that the closer such a system is to being 'natural', the more obvious will be the application of the ideas of ecological energetics.

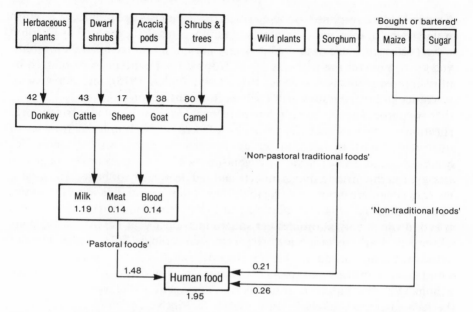

Figure 6.3   Simplified flows of energy in the Ngisonyoka Turkana ecosystem. Flows are in gigajoules (GJ) year^{-1} (Source: Coughenour *et al.*, 1985)

If we use as an example a traditional pastoralist economy from the semiarid zone of East Africa (Figure 6.3) then it is not difficult to perceive it as an analogy to the set of food chains shown in Figure 6.1; what is different is that the pastoralists live largely off domestic animals and where they engage in cultivation, cultivated plants rather than wild ones form the first level in the food web. But as both herbivore and carnivore, humans are the consumers of much of the material whose energetic pathways are plotted (notably, the study does not plot the rest of the ecosystem: it would be interesting to see what proportion of all the natural flows are intercepted by these pastoralists), and at the same time the manipulators of those flows, directing a selection of the conduits of energy-containing matter towards themselves.

Even though there may be some similarities, the realm of the human-manipulated ecosystem is different in various ways from the natural ecosystem. First, there is always the cultural dimension; nature is changed not in ways that are the product of millenia of natural selection, but by much shorter periods of selection according to the products of the human mind. This may produce results which are clearly not related deterministically to the rest of the ecosystem except at the grossest of levels. The rejections of the pig by Islam and Israel, and of the cow by Hindus, form one example; the different patterns of gardens in ancient Persia and ancient Egypt, both at similar levels of technological development, form another. Whole systems of settlement and agriculture in

China have been oriented according to the dictates of the 'way' of wind and water, in a form of geomancy.

Second, there is the capacity of human-invented technology to reach out for energy other than that provided by the sun. Early forms included the tapping of other forms of natural energy such as that of fire (stored solar energy of recent date), wind and falling water, but today the dominant source is the energy that was once trapped by photosynthesis but now comes in the concentrated form of oil, coal and natural gas. All these forms of energy can be applied to alter ecosystems (or indeed to get rid of them) in forms such as machinery, buildings and chemicals. An example of an ecological system which traps solar energy, but in which that energy is subsidized by large quantities of fossil fuel energy, is modern agriculture (Figure 6.4). Here the fixing of the solar energy upon the farm as a plant crop (which may or may not be fed to a farm animal) is supported by many flows of energy to and from the urban-industrial sector of the economy. Such is the level of support energy that a foodstuff like white, sliced, wrapped bread may have consumed two to three times the amount of energy in its production than it actually yields as nutrition when ingested (Table 6.2). So in the United Kingdom we each eat 0.53 tonnes per year of oil, and the inhabitants of the United States use more energy in their food system than the three times greater population of India use for *all* purposes.

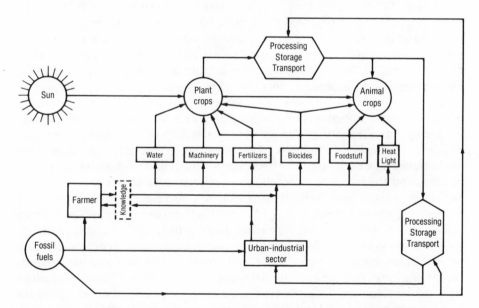

Figure 6.4 A simplified diagram of energy flows in modern agriculture. The addition of fossil fuels means that the agricultural workers can support around 32 times their own number in the urban-industrial sector. (Wastes are not shown on this diagram) (This is an original by the author)

Table 6.2   Energy budget for a white, sliced, wrapped loaf (1 kg).
United Kingdom 1970s; MJ. (Source: Leach, 1976)

Inputs	
fertilizers, biocides, machinery	
(that is, inputs to farm gate)	4.02
milling	2.68
baking	13.31
shops	0.71
total to point of sale (rounded)	20.70
Output: one loaf of 1 kg	10.6
Ratio: energy out/energy in	0.525

Note: The ratio would worsen considerably if, for example, the loaf was the
sole object of a trip by one person in a car.

In their search for and use of a higher throughput and intensity of energy than given by the natural fluxes, human societies have altered ecosystems in a myriad of ways; building roads and the vehicles that run on them, cities that radiate almost as much heat as the sun in northern winters, high-rise hotels on tropical beaches filled by tourists brought in by Boeing 747s, empty oxygen bottles on Mount Everest's summit, are all manifestations of easily available fossil fuels. These patterns have in some places been intensified by the additional supplies of electricity provided by nuclear power station, where the splitting of the atomic nucleus of uranium generates heat that can drive a turbine.

Given human behaviour, it is no surprise that some energy is put to use in destructive ways. These may often be an accidental byproduct of a particular development or an acceptable cost of a desired benefit, but they can also be deliberate as in the case of war, when ecosystems may be devastated because of the combatants' activity. In the Second World War, for example, several Pacific Islands were virtually denuded of vegetation; in the Indo-Chinese wars of liberation, the United States used defoliant herbicides to try and deny cover to their enemy, with long-term effects upon forests and mangroves (as well as people), and the intensity of bombing cratered an area 66 per cent of that of Wales (Westing, 1980). This intensity may not be needed to convert a set of ecosystems; weapons testing on the White Sands missile range in New Mexico accounts for about 1 per cent of the natural energy flux, but is enough to produce a distinctive landscape. The remnants of war in the form of unexploded mines present a problem for some forms of land use even today in places like Poland, the USSR, Libya and the Falkland Islands (Westing, 1985). The ultimate in the ecosystemic effects of the explosive release of large quantities of energy would be a medium to large-scale nuclear war. Approximately 12 000 MT of yield are available to the nuclear powers (the equivalent of 1 million Hiroshima bombs), and it has been postulated that an exchange of 5000 MT in the northern hemisphere might bring about a 'nuclear winter' with several months of darkness and subfreezing temperatures in temperate latitudes, followed by increased levels of UV-B radiation and medium-term radioactive fallout. The details of the scenarios (which are based on certain unverifiable even if reasonable assumptions about, for example, soot behaviour in the atmosphere) are not relevant here, but a return to hunter-gatherer levels of subsistences and population density over much of the planet would be a likely outcome; the extinction of *Homo sapiens* could not be excluded (Dotto, 1986). Here would be the ultimate and apocalyptic impact of human energy use upon the global system: Robert Oppenheimer's words as he watched the first trial explosion of an atomic bomb in New Mexico in 1945 were from the *Bhagavad-Gita*: 'I am become Death, the shatterer of worlds'.

This last discussion brings out the idea of a relationship between energy and power. For access to energy brings us power, and in particular power to change nature. This essay has suggested ways in which living organisms and their support systems can be changed, and there will no doubt be many others to come; would it be possible to engage in genetic engineering, for example (that is, the power to alter the basic genetic codes of living matter), without the support of energy derived from fossil fuels? But the embedded energy and mass of plants and animals are small compared with geomorphological systems such as river basins and the atmosphere. Yet human societies can change landforms quite substantially and indeed may effect an important change in the composition of the atmosphere with rather unpredictable consequences; so power over nature is a consequence of access to energy: it gives us some degree of control.

The exercise of power, whether over nature or over our fellow humans, brings us into another realm of thought. For the right to command must have a sanction and a legitimacy and this takes us from the territory of the objectively verifiable 'facts' of science (like megajoules (MJ) and the taxonomy of birds) to that of values and ethics. The title of this chapter is adapted from one of the verses (10.190) of the *Rig Veda* (Indian spiritual texts of around 1200–900 BC), 'Order and truth were born from heat as it sprang up', which can be taken as a metaphor for the linkage of human affairs with the world of nature. Clearly, energetics as a scientific study represents one of a series of successive approximations towards the truth about the order in nature, and possibly the human place within it. Moreover, we have to struggle to understand this order and what truths it may reveal in the natural world. Even further, we cannot escape the responsibility for creating in the systems which we dominate an order that reflects such truths, admittedly provisional as they must be, as we discover both about nature and about ourselves.

References

Coughenour, M. B. *et al.* (1985). Energy extraction and use in a nomadic pastoral ecosystem. *Science*, **230**, 619–625.

Cousens, S. H. (1980). A trophic continuum derived from plant material, animal size and a detritus cascade. *Journal of Theoretical Biology*, **82**, 607–618.
Cousens, S. H. (1985). Ecologists build pyramids again. *New Scientist*, **106**, 50–54.
Dotto, L. (1986). *Planet Earth in Jeopardy. The Environmental Consequences of Nuclear War*. Chichester: Wiley.
Kellman, M. C. (1980). *Plant Geography*, 2nd Edn. London and New York: Methuen.
Leach, G (1976). Industrial energy in food chains. In A. N. Duckham *et al.* (eds) *Food Production and Consumption: The Efficiency of Human Food Chains and Nutrient Cycles*. Amsterdam and Oxford: North-Holland Publishing Co., 371–382.
Lieth, H. F. H. (ed.) (1978) *Patterns of Primary Production in the Biosphere. Benchmark Papers in Ecology*, volume 8. Stroudsburg, Pennsylvania: Dowden, Hutchinson and Ross.
Lindeman, R. L. (1942). The trophic-dynamic aspect of ecology. *Ecology*, **23**, 399–418.
Odum, E. P. (1959). *Fundamentals of Ecology*, 2nd Edn. Philadelphia: Saunders.
Rigler, F. H. (1985). The concept of energy flow and nutrient flow between trophic levels. In W. H. van Dobben and R. H. Lowe-McConnell (eds). *Unifying Concepts in Ecology*. The Hague: Junk, 15–26.
Seddon, B. (1971). *Introduction to Biogeography*. London: Duckworth.
Simmons, I. G. (1979). *Biogeography: Natural and Cultural*. London: Edward Arnold.
Simmons, I. G. (1981). *Biogeographical Processes*. London: Allen and Unwin.
Stoddart, D. R. (1965). Geography and the ecological approach: the ecosystem as a geographic principle and method. *Geography*, **50**, 242–251.
Stoddart, D. R. (1986). *On Geography*, Chapter 11. Oxford: Basil Blackwell.
Stott, P. (1981). *Historical Plant Geography*. London: Allen and Unwin.
Tivy, J. (1982). *Biogeography: A Study of Plants in the Biosphere*, 2nd Edn. London and New York: Longman.
Watts, D. (1971). *Principles of Biogeography. An Introduction to the Functional Mechanisms of Ecosystems*. London: McGraw-Hill.
Westing, A. H. (1980). *Warfare in a Fragile World. Military Impact on the Human Environment*. London: Taylor and Francis for SIPRI.
Westing, A. H. (1985). *Explosive Remnants of War. Mitigating the Environmental Effects*. London and Philadelphia: Taylor and Francis, for SIPRI and UNEP.
Whittaker, R. H. and Likens, G. E. (1975). The biosphere and man. In H. Lieth and R. H. Whittaker (eds). *Primary Production of the Biosphere. Ecological Studies no. 14*, Berlin, Heidelberg and New York: Springer-Verlag, 305–328.

Energetics of Physical Environment
Edited by K. J. Gregory
©1987 John Wiley & Sons Ltd

7

Conclusion

K. J. Gregory* AND John B. Thornes†,

*Department of Geography, University of Southampton
†Department of Geography, University of Bristol

Greater emphasis placed upon the availability of energy, the use of power, and the application of energetics were suggested in the final paragraph of Chapter 1 as ways in which physical geography is now developing, and this offers a more integrated approach to the physical environment. The preceding essays have reviewed several aspects of physical geography and collecting them together has avoided the way in which such subjects have traditionally been separated as the fissiparist tendency in physical geography has increased in recent years.

Although the preceding chapters are all unified by the way in which they consider energetics in relation to some aspect of the physical environment, it is inevitable that there are differences between them. These differences explain why the subtitle for this volume is 'essays on energetic approaches to physical geography'. Not all branches of physical geography could be covered so that there is no detailed reference to mass movements, to glaciers, or to coasts despite the fact that energetic approaches have been developed in the context of these and other fields. Indeed there is some overlap between some of the preceding chapters, and the significance of water in the geosphere is a theme which appears in Chapters 1, 3, 4 and 5.

In Chapter 1 Table 1.2 indicated some ways in which five major energetic themes could be identified in physical geography research. In the light of the preceding chapters Table 7.1 provisionally attempts to demonstrate the manner in which the five themes are reflected through major spheres of physical environment. Part of Table 7.1 reflects the content of the preceding chapters, but it is emphasized that this table is a preliminary one and the reader may be able to elaborate the content as research progresses over the next few years.

Preceding chapters have illustrated that there are three major ways in which physical geographers deploy the ideas associated with energy and its transformations. These three ways can be thought of as operational, methodological, and conceptual and they are certainly not mutually exclusive although the preceding chapters have concentrated mainly on the first two.

161

Table 7.1 Examples of energetic themes in branches of physical geography

Energy aspect	Atmosphere	Sphere of physical geography		
		Geosphere and hydrosphere	Pedosphere	Biosphere
1 Sources	Internal energy Potential energy due to gravity Kinetic energy due to motion of body	Solar radiation Atomic energy Chemical energy Gravity Earth's rotation	Solar radiation Chemical energy Mechanical energy	Solar radiation Primary production
2 Circulation	Transfers, transformation, storage of energy Exchanges of heat, water, momentum Energy fluxes	Hydrological cycle Geochemical cycles	Mineral weathering Organic matter decomposition Gibbs free energy	Food chains Nutrient cycles Population dynamics
3 Budget	Energy balance	Water balance Sediment budget Erosion rates	Soil moisture budget Soil organic budget	Energy budgets at trophic levels Productivity
4 Energy relations	Available *PE* and *KE* Energy spectra and zonal balance	Force-resistance relations	Soil profile dynamics	Ecosystem dynamics Functional approach to ecology
5 Changes of distribution in time	Climatic change	Endogenetic changes		Evolution

———— Human activity ————→

Eustatic Changes

7.1 OPERATIONAL

In the operational sense, energy is considered as a basic quantity which is used to define the temporal and spatial magnitudes of events resulting from its application; the efficiencies of its transformations; and its application to the management of physical environmental systems. This can be exemplified by the energy of falling rain, the transformation of photosynthesis and the magnitude of winds blowing across the earth surface. The use of energy as the operational definition of the agency of change meets with varying success. Generally speaking, where the efficiency of energy transformation is low (compare the river with the atmosphere) then the use of energy is less successful—large applications of energy result in only small changes.

Where the sensitivity of the measurement of change is small, the energetics approach is perhaps more limited in its application than where the response is quite sensitive to small energy applications.

To further an operational energetics approach, where it is appropriate, we need to be able to express all aspects of the power of nature in similar units. This could be especially important in comparing the efficacy of human activity with that of nature, and in adopting a more coordinated approach to the control and efficiency of energy expenditure in the natural environment.

7.2 METHODOLOGICAL

In a methodological sense, energetics is used as a vehicle to organize information about physical environment into a manageable structure. The most common format here is the budget paradigm, whereby energy flows are monitored, measured, and followed through the relevant system—ecological, atmospheric or soil systems. Extended to its logical conclusion it provides a unifying approach to man and society, as in the approach of Odum and Odum (1976) to man, power and society. Alternatively a format can be structured through time or through transfers (circulation) or by identifying sources and sinks. Some of these methodological devices and their applications in the different spheres of physical environment are shown in Table 7.1.

7.3 CONCEPTUAL

There are also conceptual ways of deploying energetic approaches, when transformations of energy are used as the basis for conceptualization of processes or states based upon general principles or processes of energy expenditure. Prevalent among these is the use of the first law of thermodynamics to provide convertibility of apparently different properties to a common base. The second law of thermodynamics is used to provide principles of energy reduction and entropy maximization in closed systems. The debates about equilibrium and transcience, order and disorder, stability and instability also derive from this conceptual base. Non-linear behaviour and multiple stable states also take their origin in the debate concerning systems which, in the words of Prigogine, are far from equilibrium (Prigogine and Stengers, 1984).

From these three connected aspects of energetics, by no means an exhaustive classification of approaches, it is evident that energetics provides a rich, diverse and flexible paradigm having much in common with systems analysis, though having a sounder and more fundamental basis in the physical and biological sciences.

If an approach to physical geography embracing space, time and mesoscale approaches, without recourse to reductionism, can be underpinned by considerations of work, power, and efficiency in the language of the physical sciences, then physical geography can become more coherent and more unified. A more holistic and organismic approach, in accord with developments in many of the physical and earth sciences in recent years, might succeed the reductionist stances that have been more popular until recently. This could enable a focus upon the total environment rather than upon its components, and upon the impact of human activity and upon the most effective ways of managing the power of nature. The energetics approach has, for example, been suggested as a possible tool for analysis of environmental impact in relation to water resources planning (Meyers, 1977).

At the beginning of Chapter 1 the power of the individual and the scientific desire to harness nature were two themes identified in addition to the power of nature itself. Increased concern for the latter, through energetics of the physical environment, can only enhance the basis for appreciating the other two themes.

References

Meyers, C. D. (1977). *Energetics: Systems and Analysis with Application to Water Resources Planning and Decision-making*, US Army Engineering, Institute for Water Resources Report 77-6; Fort Belvoir, Virginia.

Odum, H. T. and Odum, E. C. (1976). *Energy Basis for Man and Nature* Wiley: New York.

Prigogine, I. and Stengers, I. (1984). *Order Out of Chaos. Man's New Dialogue with Nature*. Heinemann: London.

Energetics of Physical Environment
Edited by K. J. Gregory
©1987 John Wiley & Sons Ltd

Glossary

A number of technical terms recur throughout the preceding chapters and a series of simple definitions are proposed below.

Bioenergetics in a general sense refers to the study of energy transformations in living organisms.

Chemical energy usually refers to the energy involved in chemical reactions (chemical kinetics), and the condition or state of a reaction or of a chemical system of different reactants is therefore defined.

Cybernetics may be broadly described as the science of control in mechanical and biological systems. It is linked to energetics through the idea of information and entropy. Many control systems also operate to control the energy levels of a system or the way in which energy is partitioned.

Efficiency is the ratio of the energy applied to that which is used in doing mechanical work. Usually it refers to the ratio of work done by an engine to the heat energy supplied to the engine. In an erosional sense it is used to describe the ratio of the kinetic energy used in transporting sediment to the potential kinetic energy provided by the flow. This is the sediment transporting efficiency.

Efficiency (ecological) is the ratio between the level of energy flows at different points in the food chain (Lindemann's efficiency).

Energy may be defined as the capacity for doing work, or is that quantity which diminishes when work is done. *See also* internal, kinetic, potential, and chemical energy.

Energetics is strictly the science of the laws of energy, and is concerned with the energy transformations that occur within ecosystems.

Enthalpy is the sum of the internal energy of the system plus the product of the system's volume and the pressure exerted upon it.

Entropy is a measure of the reduction of the capacity to do work (Clausian sense). It can also be regarded as a measure of order or disorder (Shannonian sense) or as lack of information. It is expressed by the probability of occurrence of a particular state of the system.

Equilibrium in the simplest sense of a macroscopic system—one visible to the naked eye—is obtained if the system does not tend to undergo any further change. Any further change must be produced by external means. In mechanical systems equilibrium occurs when the sum of all the forces acting is equal to zero. Equilibrium between two phases (such as gas and liquid) is reached when there is no net transfer of mass or energy between the phases.

Erodibility is the degree to which the material underlying the land surface is susceptible to erosion processes.

Erosivity is the erosive energy of rainfall and runoff regime.

Flux is the rate of flow of any fluid across a given area.

Internal energy is the kinetic energy of the constituent molecules of materials. For example, when energy is transferred to a body by frictional heat generation the internal energy is increased.

Kinetic energy is the energy of motion that is the form of energy associated with the speed of an object. It is the product of half the mass multiplied by the square of the velocity and has the units joules (J).

Mechanical energy is the sum of the potential and kinetic energies of a body in motion and remains constant in conservative systems. In non-conservative systems additional work must be done to maintain the mechanical energy against frictional forces and dissipation as heat.

Potential energy is the energy of position and usually refers to gravitational energy which is that associated with the height of a body which is being acted upon by the force of gravity; it has the units joules (J).

Power is the rate of doing work. The average power provided by a process is the total work done by the process divided by the total time interval. In the SI system this has the units joules second^{-1} (J s^{-1}) which is called 1 watt (W). Sometimes it may be expressed as power expenditure per unit length (or width), for example, in relation to fluvial processes.

Synergetics is the cooperation of individual parts of a system that produces macroscopic spatial, temporal and functional structures.

Thermodynamics in its classical form is a theory which deals with the behaviour of macroscopic systems, that is systems dealt with above the level of constituent atoms. The theory depends on four laws, the *zeroth* law states that if each of two systems is in equilibrium with a third system, then they are in equilibrium with each other. The *first law* states that the work done by a system depends only upon the initial and final states even if the intervening states are not in equilibrium. The *second law* states that the entropy of an adiabatically isolated system can never decrease, and the *third law* that the entropy of any given system attains the same finite least value for every state of least energy.

Work is the change in the kinetic energy of a body as a result of a force acting to move the body through a distance. The unit of work is the work done by a unit force in moving a body a unit distance in the direction of the force. In the SI system the unit of work is 1 newton metre^{-1} (N m^{-1}), called 1 joule (J).

Index

Subjects indexed are restricted to major references. Material in tables and illustrations is indicated by page references in italics.